Memoirs

of a

Marine

Old Corps – New Corps 1919 to 1959

General Vernon E. Megee

ISBN
1-933177-28-4 (10 digit)
978-1-933177-28-1 (13 digit)

Library of Congress Control Number: 2010932262

First Edition

Printed in the United States of America
Published by Atriad Press LLC
13820 Methuen Green
Dallas, TX 75240
(972) 671-0002
www.atriadpress.com

The paper used in this book meets the minimum requirements of the American National Standard for Permanence of Paper for Printed Library Materials, ANSI/NISO Z39.48-1992.
Binding materials have been chosen for durability.

Table Of Contents

Photographs

INTRODUCTION BY
ALFRED TYSON BROAD

In February 1954 Colonel Frank Dailey, Base Commander in Kaneohe, Hawaii, ordered me to take good care of General Megee's daughter, LaVerne. He never rescinded the order. Fortunately, he chose for me the ideal wife and my best hunting and fishing partner in her father, Vernon Megee. Not only did we share ancestry in the southwestern part of our country, we were both avid outdoorsmen. We could have a great hunting day without ever firing a shot.

Around the campfire, Vernon's stories were always about flying, never about combat. The challenge of flying the Andes at above 16,000 feet in the aircraft of those days (no oxygen equipment was available at that time) were demanding, to say the least. Vernon loved that challenge and took it on without fear.

It is an honor and a pleasure to submit this additional information to Vernon's autobiography, emphasizing some interesting aspects only briefly addressed in his memoir.

Alfred Tyson Broad
Captain, USMC Reserve Retired

GENERAL VERNON E. MEGEE

In the annals of modern day ground warfare, close air support has no equal. Vernon Megee, a ground Marine turned aviator, developed the communications and tactics that allowed the ground unit commander to direct close air support for advancing Marines on Iwo Jima and Okinawa through forward observers. Until that time, control of attacking aircraft remained under the direction of a Naval Officer aboard ship, or airplanes were used to escort bombers to and from strategic targets. In subsequent battles, such as Korea, the Gulf War and Iraq, airstrikes played a vital role and the current Afghanistan operations rely on close air support tactics daily. Vernon authored most of the manuals on the subject of close air support and also served on the original faculties of the Army/Navy War College as the expert on amphibious warfare.

His tenure as Assistant Commandant/Chief of Staff to General Pate was more demanding than most knew. Although senior to, and having more combat experience than General Pate, he applied himself without pause or prejudice, having to serve much of the time as Pate's Acting Commandant in the General's absence. Following his tour as Assistant Commandant, he requested and was granted the position of Commanding General of the Fleet Marine Force of the Pacific. At his suggestion the position of Assistant Commandant / Chief of Staff was divided to the betterment of the command structure.

Ever eager to hone his skills as a marksman, Vernon set up a pistol range in the basement of the Assistant Commandant Quarters at 8th and I Streets in Washington, where he had been stationed in 1919 when he first volunteered for the corps. His wife, Nell, was content with managing the household and left running the corps to Vernon.

* * * * *

Vernon's greatest contribution to the Corps was when he became the lead spokesman, along with Commandant Clifton B. Cates, General A.A. Vandergrift, and Admiral Lewis Denfield, in the B-36 hearings before Congress in 1948. President Truman, along with Secretary of Defense Francis Johnson and the Secretary of the Navy, Lewis Matthews proposed the re-organization of our armed forces, eliminating the need for Marines except for guarding Naval facilities, and Honor Guard Ceremonies. Truman believed "If it's on the ground it's Army. If it's in the water it's Navy, and if it's in the air it's Air Force." (The air force had been established only a year earlier in 1947).

Admiral Denfield was summarily relieved of his duties because he supported the continuance of the Marines as a highly mobile, most effective amphibious force. But Congress took a firm stand, and it was proscribed by law that the Marine Corps not to be reduced to less than three divisions and three wings, with the commandant being on equal footing with the Chief of Naval Operations and the Secretary of the Navy. Vernon had put his career on the line, gaining prestige (or notoriety) that furthered his subsequent advancement, including his possibility of being the first Marine Aviator to be Commandant.

In the aftermath of the B-36 controversy, the Joint Chiefs of Staff decided to widen the membership of the staff to include a Marine Brigadier as Deputy Director for Intelligence Matters (G-3) a position then filled by Megee. He gave weekly briefings to General Omar Bradley and General George Marshall, which he considered to be one of the highlights of his military career. As a member of the Joint Intelligence Board he also sat in on weekly conferences with the director of the CIA and General "Beetle" Smith, General "Ike" Eisenhower's Chief of Staff.

* * * * *

The war in Korea proved the wisdom of Congress' rejection of Truman's efforts to get rid of the Corps. On June 24, 1950 Communist divisions of North Korea invaded South Korea. General MacArthur's

4

intelligence had failed to notice anything unusual before the invasion. The next day the United Nations called on the world to "render every assistance to Korea." On June 30, 1950, the Korean Incident became the Korean War. Commandant Cate ordered the first Marine Division to ready for war. The First Provisional Marine Brigade sailed from San Diego on July 14, 1950.

Truman and his "get rid of the Marine Corps" crowd certainly must have been thankful for their failure when the Corps landed at the Pusan perimeter. True, some of the new recruits had to learn how to field-strip their M1's onboard the transports on the way to Korea. Some of the Army contingent in the landing went ashore without any ammo at all! At least the leathernecks had ammo.

Only the Marines had continued the development of the helicopter for troop and weapon movement, medical evacuation, (MASH) reconnaissance, gunships and communications that proved invaluable in the Korean War.

"Scrape your bellies on the sand" became the by-word for fighter-bombers both Marine and Navy aviators following Vernon's orders to commander William F. Millington on Iwo Jima.

Vernon's greatest contribution to modern warfare was his development of close air support for Marines on the ground. Iwo Jima, Okinawa, Korea and all ground combat since. Many of the manuals on amphibious warfare bore his signature as author. It was also a subject he taught when he was chosen as one of three Marines to serve on the initial faculties of The National War College.

* * * * *

Vernon Megee's skill and love of flying was evidenced by his membership in the Quantico Flying Circus (predecessor to the Blue Angels) in Cleveland Ohio in 1935. He and Lewis B. "Chesty" Puller were among the 75 cadets chosen for Officer Training School in 1921. Vernon was among seventeen commissioned, but Chesty was not, *that time*. They served together frequently throughout their careers.

His love of hunting filled many local larders when he served as a ground officer in Haiti, Nicaragua and Peru, where he also served as Assistant Commandante of the Peruvian Air Force.

His command of the Spanish language and interest in local history endeared him to the local militia, residents and government officials. These skills gained him a welcoming acceptance into Lima society.

After retirement he was instrumental in the development of the Marine Military Academy in Harlingen TX. Vernon served (non-stipend) as its Superintendant. Later he became its first Superintendent Emeritus.

Alfred T. Broad, 2011

FOREWORD

I have so often been frustrated in my study of local and family history by the lack of contemporary documents and narrative memoirs that I feel justified in setting down, while there is yet time, the salient facts and impressions of my lifetime. I have written largely from memory, buttressed by such contemporary notes, letters, and official documents which have been available to me. In trying to be objective as well as accurate, I have had to pass adverse judgment on a few of my military contemporaries. Thus, I would prefer that this work not be published during my lifetime. It is written primarily as a personal record for my descendants.

To LaVerne, our daughter, and to Kathleen and Tyson Megee Broad, our grandchildren, these memoirs are dedicated with abiding affection, and a sense of family continuity that it is hoped may be transmitted to future generations.

To my wife, Nell, who has faithfully shared the adventures, pleasures and hardships, tragedies and triumphs of these forty years, I also dedicate this record of our lives together.

Vernon E. Megee
General, U.S. Marine Corps
(Retired)

Austin, Texas
1978

CHAPTER I

THE FORMATIVE YEARS

I was born in Tulsa, then Indian Territory, on June 5, 1900. My father, George Daniel Megee, and my mother, Alice Ford Megee, had immigrated to "the Territory" via the covered wagon trail shortly after their marriage in December, 1898, and were to remain there for another two years before returning to their ancestral home in Ozark County, Missouri. My earliest recollection is of the modest cottage on Uncle Paton Duggin's farm in the valley of Big Creek, a tributary of White River (now submerged under the waters of Bull Shoals Lake). I can recall the birth of my eldest sister, Opal, and certain excursions with my father, on foot or on horseback, riding behind his saddle.

Particularly fascinating to a small boy, and thus remembered, were the fish weirs that my father and Uncle Pate had constructed at the foot of one of the long blue holes of water that characterized Big Creek. The catch from these traps was an important article of our diet, particularly during the summer months when fresh meat was not to be had. According to my father's account, in later years, I persisted in following him to the fields and sometimes had to be restrained from wandering beyond my allotted bounds. It appears that wanderlust developed early!

Later we moved to my mother's ancestral farm for a season, thence to a cottage on the farm of my grandfather William Washington Megee. During these tender years, my mother taught me to read. I can recall with fond nostalgia the daily lessons held under the Sycamore tree that shaded our little house. One of my earliest books was *Lives of the Presidents*, to which I may attribute a life-long interest in

American history. At the age of five, I was enrolled in the same one-room log schoolhouse at "Stony Point" which had been attended by my father. My reading ability entitled me to "advanced standing," so that I was admitted to the third grade, passing at the end of the term to the fifth grade. I'm afraid that the scholastic standards of "Stony Point" were somewhat elastic even by contemporary standards but I enjoyed the distinction, nevertheless.

At this time, the Ozark Country had made but little material or cultural progress. The methods and implements of farming, transportation and the manner of living had not changed appreciably from those prevalent in the Appalachian region a century earlier. Ozark County had no railroad or other public transportation, and no roads worthy of the name. There were of course no telephones or other utilities, no hospitals, and no doctor nearer than Gainesville, the county seat, a day's travel to the eastward. Farm products were freighted to Springfield by horse-drawn wagons, a round trip of about ten days, and supplies for six months or a year were brought back over the flinty and rutted hill trails. These supplies were augmented as necessary, and sparingly, by purchases from the local country store, usually paid for with eggs or poultry.

Under such conditions of primitive isolation, each family and each farm had to be a self-sufficient entity. My grandfather Megee's farm was such an establishment. Timber for buildings and fences was cut and processed on the place. Corn and wheat for the miller to grind, and from which Grandmother (Sarah Futrell Megee) baked her bread, were grown in the small cleared fields along the ridge tops and in the creek bottoms. Cane for sorghum, a fruit orchard, and of course a large kitchen garden, provided for the table. Hogs that ran loose on the open range and fattened each fall on the acorns (mast) provided ample meat for the smokehouse. A few dairy cows of indeterminate breed provided milk and butter for the family, and made the evenings melodious with their bells as they wended their way home from distant hill pastures. Some cotton was grown for tax money, but a small herd of beef cattle and a few sheep, pastured on the open range, provided Grandfather with the necessary cash income.

During earlier days, Grandmother spun her own thread from homegrown wool and cotton, wove the cloth on her handmade shuttle loom, then made homespun garments for her numerous progeny. She no longer did this, within my memory, although she still kept the old spinning wheel.

The old family home was a commodious structure of hewn logs, with a gabled half-story devoted to sleeping quarters above the two large rooms below, each with a huge stone fireplace vented into a massive central chimney of rough native limestone. The kitchen was centrally located between two galleries on the south side of the main building. The roof was of heavy hand-split white oak shakes (shingles). Water had to be carried from the spring at the foot of the hill on which the house was built. Lighting was by kerosene lamps; fuel for heating and cooking came from the family woodpile.

The house, when I first knew it, was more than thirty years old, exuding a weather-beaten charm of massive durability and rustic hospitality. The furnishings were of the pioneer period, mostly hand-fashioned, bright by the rag carpets and quilted bed coverlets, which were the pride of Grandmother's hands. There were only a few books on the shelf, other than the family Bible and a Volume of Moody's Sermons. Above the fireplace, on its wooden pegs, safe from the exploring hands of a small boy, rested a long full-stocked "Kentucky" rifle, loaded and usable after more than a hundred years in the family.

The farm buildings were equally substantial of hewn logs and hand-split shingles, the corrals and small fields being fenced with split rails. All these outside improvements were the work of Grandfather and his sons. The interior trimmings were from the busy and competent hands of Grandmother and her daughters. The building and operation of such a farm represented a lifetime of prodigious labor, in which the entire family shared.

Among my early memories is the old sorghum mill down by the spring, the horned minnows and the little green frogs in the spring branch, and the May apples that grew under the trees in the little nearby flat. Remembered also is the visit of the community threshing machine, a horse-operated affair, from the vicinity of which I was banned for fear that I would "get wound up in the turnblin' rod" – the propeller shaft which ran from the horse stanchion to the machine proper.

I recall also the drowsy days of early summer with the June bugs humming among the apple blossoms. Once, I captured some, tied threads to their legs and had them making captive flights. Grandmother released them, despite my tears, and gave me a gentle scolding on the subject of cruelty.

In the early days of the homestead, there was much wild game in the surrounding hills. Many were the tales told around the fireside by my hunting elders, of bear and panther and catamount, while outside in the wintry night the hooting of the great horned owl brought delightful shivers to a small boy, and served as punctuation for the story of the wolves in Grandfather's sheepfold. Long before my time, though, the deer, the bear, the panther and the wolves had gone, and but a remnant of the once plentiful wild turkey was to be found in remote corners of the hills. There was an abundance of small game, however, which we often had on the table. Grandfather was fond of taking the big fox squirrels with his old muzzle-loading rifle. My father trapped rabbits in the snows of a hard winter to augment the family larder, while the small boy of the family found delight in tracking the cottontails through the snow behind old "Roscoe," Grandfather's venerable dog.*

Transportation over the hills was by horseback, or in horse-drawn wagons that could be heard clattering over the flinted trails for a mile or more. Buggies were seldom seen then. Occasionally, some old timer would pass with a yoke of oxen pulling his creaking wagon. My grandmother did her visiting on horseback, mounted sidesaddle, while Grandfather spent much of his time in the saddle attending to his farm business and his Sunday preaching.

Our own cottage was much more modest than the house of my grandparents and contained much less to interest a curious-minded small boy. Thus, I seem to recall being an almost daily visitor to "Grandma's house," just across the spring branch hollow from ours, where I was always warmly welcomed. The impressions of a happy fireside which were formed there were profoundly influential through my early life, serving as a living link between 'the old ways' of my ancestors and the more modern world which I was destined to enter. Due perhaps to the early death of my mother, and its domestic

---

* This same dog caused me to play truant on my very first day of school. He met me in the road on his daily visit to our house, so I felt constrained to accompany him on his rounds.

aftermath, the home of my maternal grandparents always seemed more like my own home than any other I had as a child.

The old house of hand-hewn logs was torn down in 1910, to be replaced by a more modern white-painted frame structure, which was still standing in 1978. The old log barns, now lighted by electricity, and some of the rail fences still survived to interest the fishing tourists who passed on the new paved road to Bull Shoals Lake. The old spring has been superseded by a drilled wellhead by the kitchen door, an improvement for which Grandmother would have, in her own words, "given any cow on the place."

At the time of which I write, 1905-1906, my Uncle Sam, youngest son of the family, was away attending Teacher's College at Chillicothe. My Aunt Celia, then a girl of some fourteen years, was the only child still at home. I seem to recall having been disciplined for entering her room and carrying away some colored crayons. My mother insisted that I walk all the way back across the hollow and return them, with apologies. Thus was the integrity of property rights impressed on a young mind. The lesson never left me.

My great-grandfather, Daniel White Megee, then in his eighties, also made his home with my grandparents. He augmented his Civil War pension by building ladder-back, split-bottom chairs for the local market from materials he garnered from the surrounding forest and laboriously shaped on his foot-operated "turnin' lathe." Not until his eyes dimmed with advancing age did he retire to the front porch rocker. I remember him as a sprightly old gentleman with full white beard and twinkling eye, who told us stories of the "early days" and read his Bible aloud. "Old Grandpa," as the children called him, had been a fine rifle shot. According to my father, he could outshoot any of his sons and grandsons until he was seventy years old, relying on "second sight" to aim his favorite rifle, an early original cap lock Leman muzzle-loader, which came from Tennessee. This rifle is now a family heirloom, hanging on our wall. The respect and affection with which the family treated the old patriarch is indicative of the sense of duty and responsibility toward the elders of the clan that was the rule in that pioneer society. There were no "Old Folks Homes" in the Ozarks.

My other great-grandfather, Bryant Futrell, who lived but a few

13

miles away, has not left much of an impression on me. According to my father, who was one of his favorites, "Uncle Bry" was a frugal, industrious "Frenchman," with an appreciation for the fruits of his vineyard. I remember going to his funeral, riding behind my father's saddle across the hills to the home of Great Uncle Henry Futrell. It was not the solemn ritual which impressed me, however; rather it was the prevalence of large rats in Uncle Henry's barn, and the fun the younger kinsmen had hunting them down with a black spaniel. I feel that great-grandfather Futrell would have approved this juvenile version of an Irish wake!

The Futrells, an old North Carolina colonial family, had come to the Ozarks immediately following the Civil War, after an interim sojourn of more than two generations in Kentucky and Tennessee. Grandmother's immigrant ancestor, Thomas Futrell, had come to "Virjiny" from England during the latter half of the seventeenth century. Her great-grandfather, Nathan Futrell, served as a drummer boy at the tender age of seven during the latter part of the Revolutionary War, thus establishing DAR credentials for his female descendants.

Grandmother well remembered their migration to Missouri, especially the crossing of the wide Mississippi River at the mouth of the Ohio, and the long overland trek by ox wagon. She could also recall the marching soldiers of both armies, and the distant booming of cannon during Grant's attack on Fort Donelson.

My mother's parents, William B. and Lou Stover Ford, died long before I was born, so that I never knew these other grandparents, and after my mother's death, we had but little contact with the scattered members of the Ford family. My great-grandfather, John Ford, and my grandmother, Susan Graham Ford, had come to the Missouri Ozarks from Kentucky about 1818, thus ante dating by more than forty years the arrival there of the Megee clan.

In the summer of 1906, following the birth of my younger sister, Walsa Pearl, my father yielded to the blandishments of his older brother, Tom, who was about to set out on another of his covered wagon jaunts which had already taken him and his family deep into Texas and through most of the Indian Territory, where former neighbors had preceded us, and whose glowing reports of rich cotton lands made the flint hills of Ozark County seem doubly sterile.

Early in August, the caravan of three covered wagons was assembled in Grandfather's barnyard for the final departure. I recall little of the sadness of the occasion, except Grandmother putting her apron to her eyes as we rumbled down the lane. I was unhappy because I had to leave my little red wagon behind, and no doubt excited at the prospect of strange adventure ahead. It would be twelve years before I was to visit the old homestead again. The journey westward consumed some three weeks over the unimproved roads and trails of the day, and was one long memorable adventure for the children of the two families.

The first train was encountered near Sparta, Missouri; the first automobile, which frightened our teams almost out of their harness, was met on a hilly road near Neosho, and the first-remembered Indians and Negroes were seen on the streets of the little cow town of Claremore, in the Territory. We camped at night and "nooned" by day in some handy grove along the roadside. Cooking was done over open fires. The women folk and the small children slept in the wagons, the men and boys on the ground. Our horizons expanded daily; the well-equipped farms of western Missouri and northeastern Oklahoma (then Indian Territory) were a source of wonder to children accustomed only to hand scythes and bull tongue plows.

Even so commonplace an article as a well pump enticed two small boys to enter a strange yard and start pumping water, to the consternation of their parents, and the probable amusement of the pump owner. The journey ended at last, in the sweltering dog days of late August, when Father found a home for his family on a prosperous cotton farm south-east of Chandler, in what is now Lincoln County, Oklahoma.

The farm owner and his family were kindly, cultured people, who had homesteaded there in 1889, built an imposing home and developed a fine farm. They needed assistance in their operations, so made us welcome in a one-room cabin on the premises until a new house could be built for us. Here we spent the autumn and early winter months while Father helped with the cotton harvest. I recall the treasury of books and magazines that were made available to me as soon as it became known that I could read at that level. I was entered in Morning

Star School, being demoted to the third grade as being more commensurate with my age of six years. This comparatively modern country school, with two classrooms and fine equipment made a lasting impression on the boy who had started in "the little log schoolhouse" deep in the Missouri Ozarks.

After Christmas, my mother became ill of a lung congestion, weakened steadily, and passed away on January 31, 1907. This was for the small boy and his young sisters a stark personal tragedy, a loss rendered even more acute by events to follow, and which profoundly influenced their lives throughout childhood and into maturity. My grandparents came out immediately, bringing with them "Old Grandpa" and my young aunt, Celia. They moved into the new house which had been planned for us before mother's death, and stayed there with us for several months. Grandfather was often my companion afield, teaching me to handle a horse and demonstrating his many useful skills. Grandmother took us to her ample bosom, doing what she could to fill the aching void. I recall that one of my chores was to see that my great-grandfather, then approaching senility, did not get lost on his daily walks. I remember little, actually, of my mother as a physical person, except that she was petite, with auburn hair and large brown eyes. Her interest in my continuing education was predominant in our relationship, indicative of her character and ambition for a better life for her children. The doors of learning which she opened so alluringly at such an impressionable age fostered in her son an avid love of reading and an inquiring mind which were destined to carry him on to accomplishments of which only his mother dreamed. Truly, in her tragically short life, Alice Ford Megee accomplished much more than she could possibly have realized.

My father, in July of that same year, announced his forthcoming marriage to a widow with two sons younger than I, which failed to elicit any enthusiasm from the family; particularly on the part of grandmother, who could well foresee the difficulties inherent in a mixed family. My grandparents moved into a neighboring house long enough to permit harvesting the crop which grandfather had planted that spring, then returned to their old home in the Ozarks, taking with them my baby sister for what proved to be a two-year visit. For me this was the end of a happy era. I was not to see my grandparents for nearly eight years.

Adjustment to the new stepmother was slow and often painful. She proved to be an industrious, competent housewife, who provided well for the material needs of her new family. Unfortunately, her spiritual qualities were not equal to the task of replacing, in the affections of her stepchildren, the real mother whom they had lost.

This was particularly true for the seven-year-old boy, who best remembered his mother and who never could bring himself to accept the substitute. Thus were implanted the germs of animosity and rebellion which cast such a shadow over the remaining childhood years; and which without doubt carried over into maturity as unfavorable personality characteristics. While in later years a tolerable relationship developed between my stepmother and myself, it would be less than honest to pretend that we were ever compatible. The fault, as usual, might be said to rest with both sides. The situation was unfortunate for both of us; the scars have never quite healed.

In November 1907, Oklahoma and Indian Territories were united and admitted to the Union as the State of Oklahoma. The local celebrations were noisy and exciting to a small boy, who remembers that his elders approved the closing of the open saloons in what had been Oklahoma Territory.

The following spring brought the big flood that raised the nearby Deep Fork of the North Canadian River to awesome proportions. Father took me to see the mile-wide waters, the swimming cattle, and the occasional cabin or barn being borne downstream with the driftwood. That same year brought rumors of the dreaded "Night Riders," who preyed on late travelers and isolated households. Farmers carried Winchesters in their cotton wagons, and tried to avoid late turns at the local gins.

In January 1909, we moved to a leased farm in the neighboring "Stone School" district, which was to be our home for the next eight years. By this time, I was large enough to lend a helping hand with the farm work, and eagerly accompanied Father to the fields. He was always kind and patient in teaching me his skills and by the time I was ten-years-old I could handle a gentle team of horses and the lighter farm implements. My education continued apace. I was reading, indiscriminately, all the books I could get my hands on. Some of these,

I recall, dealt with the Spanish-American and the Civil Wars, sparking my first interest in things military. There was also Stanley's Darkest Africa to whet my taste for adventure, and Horatio Alger's works to kindle my ambition.

We attended school eight months out of the year, from late November to mid-May, then during July and August for the summer session. The split term was designed for the convenience of the cotton farmers of the district whose children were needed to work in the fields during chopping and picking seasons. There were no school buses. We walked our two miles to school morning and afternoon, regardless of the weather. Sometimes we actually suffered from the cold, and in the summer, the road was long, hot and dusty. In the spring the roadside and the adjacent meadows were a mass of wild flowers, which we eagerly awaited, and could name one-by-one as they emerged. Whatever the season or the weather, however, we never considered our daily hikes as any hardship, quite the contrary, in fact. Very rarely, and then only under the most severe conditions of cold or heavy rain, did Father ever leave his work to drive us to or from school.

In retrospect, the quality of our instruction was excellent. We were exposed, under competent supervision, to the fundamentals of education minus the frills. For organized diversion, we had school plays, Christmas, and graduation programs, for which the community turned out in force. We had no PTA organization as such, although I recall that our parents always took a very active interest in our community school.

In that era of Central Oklahoma, cotton was the cash crop and the mainstay of the local economy. It was the custom on leased farms to plant two-thirds of the arable land to cotton each year, a practice that eventually ruined the land through exhaustion and erosion. Cotton production demanded a prodigious amount of hard labor that, if hired at going rates, cut deeply into the possible profit. Thus, the most successful cotton growers were those with the largest families who could take the field, regardless of sex, during the peak labor seasons. I'm afraid that Father, with but one small son to help him, did not find his endeavors at that stage too profitable.

Each year, though, as I grew larger and stronger, my share of the farm load increased in proportion. At twelve, I could handle a plow; at fourteen I was a thoroughly competent farm hand. From that time onward I was considered entirely self-supporting, an asset instead of an economic liability to my harassed father. I was born with a love of the land, and the hard physical labor of my youth never altered my interest in agriculture and country life.

Although by the standards of later generations, my youthful leisure hours were severely circumscribed, yet I retain pleasant memories of being often afield and in the woods with dog and gun, or along the streams with an improvised fishing pole. My father taught me to handle a gun early. At ten, I had my first .22 rifle, and then my own shotgun at fourteen. In the winter months, I ran my trap line before and after school hours, supplying myself with spending money from the furs of 'possum, muskrat, mink and skunk, which I gathered and sold. I never had an allowance, as such; Father had his hands full supporting the five younger members of the family.

My companions were mostly my numerous cousins, progeny of Uncle Tom and Aunt Mary, who lived on an adjoining farm. Theirs was a noisy, happy family, and I was often at their house. Uncle Tom was a gun fancier, and he could tell interesting tales of his boyhood days in the Ozarks and his later adventures down in South Texas and in the "Choctaw Nation." Aunt Mary's father, "Uncle Bob" Shaw, was a Confederate veteran who still carried a Minie' ball in the calf of his leg, and somber memories of his sojourn in a Yankee prison camp. Other old timers in the neighborhood were Union veterans. Some had been Indian fighters. Their reminiscences in pungent and picturesque language, which my religious father often thought unfit for my ears, gave me a first-hand insight into American history.

Current happenings of the period, such as Chief Crazy Snake's abortive Indian rebellion of 1907, and Henry Starr's daring noonday robbery of both banks in the neighboring town of Stroud, on March 27, 1915, added to my historical perspective. Starr was shot off his horse by a sixteen-year-old boy, Paul Curry, who also winged another of the bandits with his Winchester as they rode out of town. Their ensuing trial, which I attended in part, was my first experience with court procedures and the working of American justice.

The sheriff of our county, during part of this period, was Bill

Tilghman, famous frontier marshal and one-time terror of the lawless element in Dodge City. I remember him as a mild-appearing man of medium height, with drooping white moustaches, who could often be seen sitting outside the courthouse basement, which contained his office and the county jail. Although he was reputed to have been one of the fastest men with a gun who ever helped to tame the old West, I never saw him wearing one. He was never too busy to pass the time of day with the admiring small fry who came by to see him.

By the time I was twelve years old I had practically completed the grade school curriculum, but kept on attending classes for two more terms because there was no other outlet for my pent up ambitions. At fourteen I was formally certified as having completed the eighth grade level, and in the normal course of events should have gone on to high school. Unfortunately, the way was not open to me. The nearest high school was located in Chandler, three miles away, requiring attendance throughout a regular nine months term, an arrangement which gave no thought to the convenience of a farm boy. My father did not feel that he could spare my services from the farm. Had my mother lived she would have found a way. So I reentered the community school for another year in "graduate status," where understanding teachers undertook to tutor me in some of the high school subjects. My passion for reading now embraced such periodicals as *The American Boy*, the *Youth's Companion*, *Hunter – Trader – Trapper*, *The Path-finder* for current events, and *The Saturday Evening Post* for fiction. I did not escape exposure to the more lurid dime novels of the day, which glorified "Buffalo Bill" and the "Old West." Their influence, if any, was counteracted by the works of Stewart Edward White, Zane Grey, and a serious tome by Gifford Pinchot, entitled *The Training of a Forester*.

In the summer of 1916, I was permitted to accompany two of my older cousins on a train trip to Southwestern Oklahoma, which was my first acquaintance with the Wichita Mountain country and old Fort Sill. We worked in the grain harvest for two or three weeks to defray our expenses and develop new muscles, before returning home. Later in that same summer, probably to escape what I had come to consider intolerable home conditions, I made another train trip to Eastern Oklahoma, to visit my Aunt Rana Gaulding and my Uncle Sam

Megee, both of who lived then at Checotah. Uncle Jim Gaulding owned a grocery store in town, and my cousin, Don, was the delivery boy. Of course, I lent them a hand, gaining some new experiences along the way. Uncle Sam taught in one of the adjacent country schools, commuting by horse and buggy. I found it interesting to accompany him, "visiting school" and soaking up his philosophy of life. As the educated elder of the family, my Uncle Sam was always my valued counselor, insisting that I must continue my education, by some means or other. He it was who persuaded me that I should return to my father's home and help him harvest his crops. This I did, after a short interlude as a hired cotton picker on a neighboring farm.

During the summer, my father had decided to join his brother, Tom, who had moved to Kiowa County. Thus, in early November 1916, we again uprooted ourselves and took to the covered wagon trail as somewhat belated pioneers. This time the family traveled by train, while Father and I and Jeff, the hired man, drove the two wagons. This was a pleasant adventure for a sixteen-year-old boy. The autumn weather was delightful, the trip across the prairies to Chickasha, up the Washita River through Anadarko, Fort Cobb, Carnegie and Gotebo, thence south around the west end of the Wichita range to Mountain Park, was a leisurely one consuming some nine or ten days. There was ample time to study the countryside and enjoy the roadside hunting.

The year spent at Mountain Park as lessees of an alfalfa farm on East Otter Creek was a busy one, with no time for school. The hunting and trapping along Otter Creek was excellent during the winter. I managed my first bicycle, and thereafter did considerable exploring of the surrounding countryside, going as far afield as the Wichita Mountains to the eastward and the Salt Fork of the Red River to the west. The winds and the dust storms ushered in a rainless spring, which dragged uncomfortably into a summer of burning drought. The upland fields and pastures became a literal desert; only the sub-irrigated alfalfa continued to produce a lush cutting of hay each month. Sold at wartime prices, the alfalfa crop saved the day for us economically.

During the spring we enjoyed a memorable visit from the grandparents whom I hadn't seen since 1907, and wasn't quite sure that I could recognize when Father sent me to meet them at Snyder.

They stayed with us for a month. Grandfather spelled me at the plow, remarking on the tremendous size of our fields and the baked hardness of our soil. He was invited to preach one Sunday at our church in town. I attended with some reservation, wondering if this Ozark circuit rider would impress our more sophisticated town congregation. I was relieved and delighted with his erudite delivery, and not a little ashamed of my doubts.

By late summer, Father and family had had enough of the "dust bowl," sold out the lease and returned to Lincoln County. I made the trip back to Oklahoma City with a companion on our bicycles, a three-day adventure which we accomplished on five dollars each and some ingenuity. Father chartered a freight car for his stock and implements, returning the family by train. Pioneering, as such, had ended for us.

At this point I began to chart my own course, working at various jobs until November, when I entered Oklahoma A and M College at Stillwater as a provisional student in the School of Agriculture, and of course a member of the military cadet Corps. At long last, I was free to pursue my interrupted formal education, and I made the most of the opportunity. During the next summer, I helped Father briefly with his farm chores, then left with my companion of the bicycle adventure, Bill Brown, for the Kansas wheat fields where the shortage of harvest hands due to the war had become acute. The wheat farmers were offering the unheard of wages of five dollars a day and board. This seemed a golden opportunity to finance my next school year. We were gone two months, returned with hard muscles, a second-hand motorcycle, and empty pockets. Our fathers were not impressed.

I stayed home that fall and helped Father with his cotton harvest. The whole family, excepting myself, was stricken with Spanish influenza, and Father was incapacitated for weeks. I had to assume adult responsibilities as acting head of the family, a sobering experience. Earlier, before the influenza struck, I managed to get away long enough to visit my grandparents on the old Ozark homestead, in those days a three-day trip by train, "mail hack," horseback or on foot. After several delightful days visiting relatives and hunting across the colorful hills, Grandfather rode with me to the forks of the road, from whence I was to take my departure for Lead Hill, Zinc, and the Missouri Pacific Railroad, which by a devious route was to carry me home. It was wartime, and I expected soon to be in uniform. I can still see Grandfather, sitting tall in the saddle as he bade me farewell with some homely words of advice, showing little evidence of his sixty-eight years, dignified, righteous, and kindly, a grandfather to be respected and cherished, an example to be emulated. I never saw him again.

I returned to college after Christmas, but a liking for military life gained as a member of the R.O.T.C., and a strained economic situation, combined to cut my second term short. After an interview with the college president, who kindly expressed his regret, I was granted an honorable discharge in late February 1919, and some two weeks later began my long career as a United States Marine. This rather precipitate action proved a terrible blow for my father, but could hardly have been entirely unexpected, considering the situation long existing in his household. Perhaps largely because of it, my father and I were always very close, and continued to be so throughout his long lifetime. He was a gentle, kindly man, beloved of his children, caught up in the currents of life with which he was sometimes unable to effectually cope. His religion was genuine, the staff of faith, which sustained him during the crises of his life, and was a powerful factor in the formulation of his children's characters.

Biographical details are of general interest only inasmuch as they interpret character and personality. My childhood experiences, though often unhappy, helped to put iron in my soul. Money was hard to come by; thus I learned early the dignity of labor, the value of thrift, and the inconvenience of being without funds. Education was valued in direct proportion to the difficulty of obtaining it; I valued it most highly. The common virtues of truth and integrity were not left to heredity; my elders did not fail to press them on my attention on every conceivable occasion. While I have sometimes felt that I was given an overdose of religion, the precepts of Christianity have ever helped me in the development and maintenance of acceptable moral standards. In summary, the somewhat unusual background of my youth disclosed to me a first-hand panorama of America's development, from the oxcart to the airplane; and an appreciation of the homely virtues, which made such progress possible. It seems to me, in retrospect, that I had a solid foundation on which to build.

CHAPTER II

TRANSITION

Oklahoma City, in late February of 1919, was swarming with uniforms. The Armistice had been in effect for more than three months, and the boys were coming home from the cantonments, the aerodromes, and the ships at sea. They were interested only in civilian clothes and jobs; recruiting stations were definitely not popular.

The Marine sergeant in his smartly creased forestry greens was a dapper individual, made somewhat sinister in appearance by the wound scar on his cheek. He was of the old bReederer, the beribboned professional who had no interest in demobilization; his job was to find new recruits. He welcomed the diffident applicant affably, did his selling job competently, and probably complimented himself on his persuasive powers. The applicant, keeping his own counsel, did not reveal that he had previously read the prospectus, that he was convinced that the Marine Corps offered superior advantages to a young man seeking to improve his status, and that his mind was already made up before he had set foot in the recruiter's lair.

In any event, he was put aboard the night train for Kansas City, and in due time found himself in the Marine Recruiting Station, standing at accustomed attention in the presence of the recruiting officer, remembered as a first lieutenant of mature years and dignity, a long-time first sergeant with a war time commission, whose discerning first question had to do with previous military service. Satisfied with the R.O.T.C. explanation, he handed the applicant some papers, some railroad tickets, and the custody of four other young men, and had his sergeant escort them to the night train for Memphis, Atlanta, Augusta,

Yemassee Junction, and Port Royal, South Carolina. This trip required some forty-eight hours of day coach travel, recruits not being entitled to the amenities of George Pullman. Since this was the beginning of the big adventure, there were no complaints.

There came aboard at Augusta, as fellow passengers, a formidable trio of hash-marked old timers fresh from the China Station, now enroute to the Marine Corps Recruit Depot, Parris Island, South Carolina, to drill the post-war recruits. After an interested appraisal one of the applicants ventured to engage them in conversation, and was rewarded by several hours of lurid reminiscences, which opened hitherto unsuspected vistas of military life.

At Yemassee Junction, where the Atlantic Coast Line Railroad cuts through the swamps on its way to Florida, a tough-visaged Marine sergeant wearing a duty belt, unceremoniously rounded up the applicants from the West, added them to others he had corralled, brusquely pushed the voluble old timers into another coach, and thus initiated the prospective recruits into the mysteries of military segregation.

The decrepit pride of the Charleston and Western Carolina Railroad screeched to a stop on the dock at Port Royal. The duty sergeant mustered his rather nondescript charges in the gathering darkness and marched them aboard a fussy little tugboat. Within the hour, they had landed on the Parris Island dock, endured the amused taunts of the lounging marines, and had been marched over a mile of oyster shell road to the Receiving Station, the Ellis Island of the establishment. Here they were herded into a mess hall for a warmed-over supper, and then handed dishtowels. The initiation into the fraternity of Marines had formally begun.

Such was my first step in the transition from civilian to soldier. There were to be other and more difficult passages to negotiate before that metamorphosis was complete. Even today, I wince a little at the memory of those next two months.

Recruit training within the Marine Corps has long been based on the principle of shock. Once he has taken the oath of enlistment, the aspiring candidate is propelled through a gantlet of indignation calculated to put him in a submissive and receptive frame of mind at the earliest possible moment. This system was originally designed for

25

the lowest common denominator, the thickheaded ignoramus who knew nothing of disciplined life, and whose small brain cage and lack of muscular coordination drove drill sergeants to insane fury, simulated or real. There were no exceptions to the harsh rules of conduct profanely prescribed by these minions of the established order. College lads who found themselves caught up in this system suspended their intellectual functions for the first two or three weeks, until they could devise ways and means of survival with a modicum of dignity and self-respect.

The only part of recruit training which I really enjoyed was the two weeks spent on the rifle range. This was where an affinity for arms attracted grudging attention from my platoon sergeant, an educated young man enlisted for the "duration of the war," and who resented being kept on to drill what he contemptuously referred to as "after the war heroes." The fact that he had never been closer to France than the dock at Port Royal, having spent his entire service drilling recruits, did not improve his cynical and sadistic humor.

The spring of 1919 was, in fact, a period of disintegration for the Marine Corps. From a wartime peak of some 75,000 men the Corps was rapidly shrinking to pre-war dimensions. The duration enlistees and the reserve officers were leaving; the regulars who had won accelerated promotion were now undergoing deflation. Majors were reverting to captain rank, captains to lieutenants, or to warrant grade if they had come up from the ranks; lieutenants and warrant officers, originally sergeants, were sewing their stripes on again. Everyone seemed to be disgruntled, understandably so, and there appeared to be little professional attention devoted to the training of recruits. We seldom saw an officer, except for those few salty characters who had themselves been drill sergeants, such as Captain Jimmy Wayt and the Borgstrom brothers. The Recruit Depot merely coasted from its wartime impetus until the slack would again be taken up.

It was not a very propitious time for a young man to be entering upon a professional military career. The officer's training schools had been suspended, as had the wartime non-commissioned officer schools. There was no indication when this formal training would be resumed, if ever. I do not recall, however, that this rather dismal prospect caused me any great concern, "sufficient unto the day was the evil thereof!"

The strenuous training period ended at long last, and the recruits graduated to Marines. The members of my company were scattered to the various posts of the Corps as replacements for "duration" men. I, who had dreamed perhaps of some glamorous foreign duty station, found myself as a clerk in the Post Quartermaster's officer, a victim of what then passed as career management. This unwelcome arrangement I shortly escaped by admitting to some familiarity with internal combustion engines; whereupon a harassed and short-handed motor transport officer, one Captain Charlie Baylis, placed me in charge of the one-man motorcycle repair shop. I remained with the Post Motor Transport Detachment as a mechanic, a truck driver, and as a dispatcher for the next year, despite my calculated efforts to escape the confines of Parris Island.

In retrospect, however, life during these months was not unpleasant. I took my first furlough in August and returned home, rather proud of my blue uniform, despite its obvious unsuitability to the summer climate of Oklahoma. There was later an excursion to Savannah on the paddle wheel steamer, which served the rivers and estuaries of the coastal country. There were exploratory hikes and fishing trips, an occasional trip to Beaufort, even one weekend hunting trip to adjacent St. Helena Island. There was the Post Library, and the organized evening classes conducted by the local Y.M.C.A. The Post Marines enjoyed considerable liberty of action compared with the recruits, who enjoyed none.

During the summer of 1919, the activities of Marine Corps Aviation, formerly at Miami, Florida, were transferred to Parris Island. The first contingent of two "Jennies" and their crews were housed in tents on the West Wing parade ground. An exhibition flight, scheduled for the Fourth of July celebration, came to grief when the pilot failed to pull out of a falling leaf maneuver and crashed head-on into a wooden barracks, fortunately used by the Quartermaster as a storehouse for mattresses! We pulled the two aviators out of the wreckage, somewhat battered but alive, and were duly impressed with their intrepidity, and the hazards of flight in the stick and wire machines of the day. The old rifle range at the Receiving Station was converted into a flying field, hangers were built, and the rest of the squadron came up from Miami. What I seem to remember most was the frequency of accidents, some of them fatal, and the devil-may-care attitude of the young aviators.

In the spring of 1920, after qualifying on the range as an "Expert Rifleman" and pinning on the coveted silver badge with its crossed rifles and wreath, I was entered in the South-Eastern Division rifle matches as a local competitor. I recall that the range officers were Captain Joe Jackson and Gunner Calvin Lloyd, famous old-team shooters of the Marine Corps, and that the chief coaches were non-commissioned shooters of the same stripe. I should like to recall that I won the match, or at least placed for one of the lower ranking awards. In all honestly however, I must record the sad fact that, having done well at short and medium ranges, I used up seven of my twenty shots searching for the target on the fog-shrouded thousand-yard range. Thus, was I introduced to the science of doping wind and weather, at which the old-team shooters were so adept, and to which no neophyte need aspire in his first season.

In May of that year, 1920, there was a diplomatic flare up with Mexico, and a punitive expedition was hastily organized. We formed a battalion from Post troops and the half-trained recruit companies, entrained for Charleston Navy Yard, boarded the Navel transport, *Henderson*, and set sail for Philadelphia to pick up the rest of the provisional Sixteenth Regiment and to load supplies. My job during the loading was to uncrate and assemble a number of new Indian motorcycles and see that they were loaded aboard for instant use on debarkation. After some feverish days the expeditionary force was assembled, and the *Henderson* steamed down the River, past the Delaware capes and out to sea on a southerly course. Several crowded days later we dropped anchor off Key West, loaded supplies, then put to sea again, not for the Mexican coast, as we had expected, but for Pensacola, Florida, where the regiment went under canvas outside the walls of old Fort Barrancas and undertook some "shake down" training. Here, in recognition of my assignment as "acting chief mechanic" for the regimental motor transport, I was promoted to the rank of corporal (temporary).

As the flaming summer dragged on, the diplomatic crisis eased; the *Henderson* took us aboard again and set course for the West Indies.

We stopped at Guantanamo, Port au Prince, Santo Domingo, and the Virgin Islands, exotic places to the young men who had never been outside the States before. At each port we exchanged our fresh-faced recruits for sallow veterans of the malarial belt who had "fought the way" in the Caribbean outposts, and who were prone to express their frustration in indelicate terms. Association with these repatriates during the voyage home helped to dispel the Richard Harding Davis myth of romantic soldiering in the tropics.

The regiment was disbanded in Philadelphia, after a short life of some two months. My detachment was returned by rail to Parris Island, and those of us who had enjoyed temporary exalted status laid aside our stripes and picked up our wrenches. It had been a fascinating interlude; we were now expeditionary veterans with something to talk about, the balmy blue waters of the Gulf Stream, the flying fishes, the storm off Cape Hatteras, the Marine sentry who fell overboard and was promptly rescued by a small boat crew. Forgotten were the cramped and malodorous quarters, the necessary regimentation of all functions, and the enforced idleness inseparable from extended troop cruises. Remembered were the flaming sunsets in the Gulf of Mexico, the distant blue hills of the Cuban shoreline, the diurnal passage of myriads of fishing boats in the Gulf of Gonaives, and the rusted bones of the cruiser *Memphis* on the reef in the harbor of Santo Domingo.

Some months later, I was selected to attend the U. S. Army Motor Transport School, at Camp Holabird near Baltimore, Maryland. This was in effect a consolation prize since I had requested assignment to the Naval Aviation Mechanics School at the Great Lakes Training Station, following my first flight in a twin-engine flying boat while we were in Pensacola. I accepted the substitution with good grace, as I recall, and spent a very interesting and rewarding six months, learning, under competent civilian instructors, the more advanced phases of auto mechanics. I returned to Parris Island in April 1921, with enhanced mechanical skills, prepared for a prolonged stay as payment for my education.

Fortunately, this was not to be. Major General John A. LeJeune, then Commandant of the Marine Corps, had shaken up a somnolent Headquarters and announced some new policies. Among them was the resumption of the limited pre-war practice of offering regular commissions to "meritorious non-commissioned officers." The policy, as expressed in a Marine Corps Order, was "to keep the door open, but to carefully select those who should pass through." To implement this program a "Candidate's School" was to be convened at the Marine Barracks, Washington, DC, in June, 1921, of six months duration, at the end of which the survivors would be given an academic examination to determine their fitness for commissioned rank. Recommendations were immediately solicited from commanding officers.

While I had originally enlisted with some hope of eventually winning a commission, the post war deflation had disillusioned me, and I had almost given up hope. I read the order with interest, certainly, but since I was still rated as a private, first class, I didn't consider myself qualified under terms of the order. My immediate commanding officer, of respected memory, Quartermaster Clerk William J. Gray, who had served as a commissioned officer during the war, fortunately for me, felt differently. After questioning me at length, he arranged for a special promotion to corporal; a letter of recommendation to the Commandant followed, and within the month, I had received my notice of acceptance and was in Washington.

The "Candidate's School" was designed to prepare the aspirants to pass a written examination in English, Literature, History, Government, and Mathematics, at about second-year college level. The curriculum also included military drills, ceremonies, and minor infantry tactics, following closely the practice at the Service Academies. Discipline for the candidates was rigid, study hours were long, and there was little time to enjoy the diversions of the capital city. Class standing was deemed most important, and competition was very keen. Many of the seventy-five who started the course were eliminated for one reason or another. Some forty took the final examination; seventeen passed. I was among the fortunate ones, standing half way down the list in point of academic merit. Following graduation, we were given short leaves to visit our homes while our commissions were being processed.[*]

---

[*] While undergoing instruction in late 1921, my class of officer-candidates participated in the formal ceremonies incident to the interment in Arlington National Cemetery of the Unknown Soldier.

One of the successful candidates had been a member of the Naval Academy class of 1922, due to graduate in June. According to existing law, it was held that he could not be commissioned until after his Naval Academy class had graduated. The Commandant undertook to get the law changed; meanwhile, for some administrative reason, holding up all our commissions. We were nevertheless packed off to the Marine Base at Quantico, Virginia, early in February 1922, to enter the newly formed Officer's Basic School, which was designed to train us in our specific duties as Marine lieutenants. Some of the instructors treated us as officers, as they should have; others were not so gracious to students who still wore their enlisted uniforms. This awkward situation endured for some three months. Finally, the legislative barrier was overcome, and our commissions were delivered, dated back to February for purpose of rank and precedence, but carrying no provision for the difference in pay that we had suffered.

We were sworn-in as officers on May 5, one month before my twenty-second birthday. Our new uniforms had been ordered long since, so we lost no time in assuming our new prerogatives. By nightfall we had returned the salutes of half the Marines in Quantico, and within the week we considered ourselves to the manner born, possibly because we had so long anticipated the occasion.

In those days, the gulf between enlisted men and officers was wider than it has since become. To pass over to the commissioned side was indeed a difficult accomplishment, involving the acceptance of an entirely new set of obligations and social values, and a complete severance of all former personal associations. None understood this better than our erstwhile companions who remained in non-commissioned status, and who scrupulously observed the military formalities whenever we afterwards chanced to meet. We accepted our new role with what I hope was proper humility, and took particular pains to conform in all respects to the traditions, customs and regulations, which had governed the conduct of Marine officers since 1775. We realized that we were being critically appraised by our brother officers, some of whom were slow to accept us into their inner circles. Our position, in this respect, was not enhanced by the presence within the officer's corps of a few ex-sergeants of long service, whose personal habits and lack of the social amenities did not match the war records which had obtained for them their permanent commissions.

We were the first post-war class to avail ourselves of this route to commissioned status; other classes were to follow, until the pattern of selecting annually from the non-commissioned ranks approximately one-third of the new officers to be inducted, had become firmly established and universally accepted. In practice, the candidates for commission were almost invariably selected from the junior non-commissioned officers, serving in their first enlistment, with some record of college-level education. The requirement of a stiff academic examination, and the age limitation of twenty-seven years, effectively barred the door, during peacetime, to the mature non-commissioned officer of long service but short formal education. Consequently, the officers appointed from the ranks during the period between World Wars were comparable in age and background to their contemporaries appointed from the Naval Academy and the list of "distinguished military colleges." While most of them lacked a college degree, their service experience compensated for this lack insofar as their professional competence was concerned. In later years, it would have been very difficult to distinguish among Marine officers of any year group with respect to origin alone.

This broadening of the base for the Marine officer's corps has had salutary results, and has been credited by many authorities as being partly responsible for the enviable *esprit de corps* in battle, which has characterized the well-led Marines. Those of us who pioneered this innovation may well reflect, with considerable pride, on the part we played in this development over the years.

CHAPTER III

PREPARATION

Quantico: 1922 – 1923

By June of 1922, the Marine Corps had completed the transition to peacetime status. The officer corps had survived the psychological trauma induced by the action of the Russell Board, which demoted the temporary officers in an effort to restore the *status quo ante bellum*; and the corrective action of the subsequent Neville Board, which not only restored these officers to their wartime ranks but confirmed them in permanent status. These boards, which followed each other within a few months, reflected the divergent philosophies of those senior officers, like Colonel John H. Russell, who had not served in France, as opposed to those like Brigadier General Wendell C. Neville, who had. The Neville faction prevailed; thus the Marine Corps was destined to be ruled for the next generation by those officers who had served in France with the second Army Division.

To compound the confusion of the times, the Snyder Board was convened to determine the relative ranks of those junior officers who had not had the opportunity to distinguish themselves during the war. After lengthy deliberation, and as a final gesture of brilliant logic, one group of some fifty ex-reservists had their names placed on the permanent list of Marine Corps officers strictly in alphabetical order. Thus, Alburger outranked Zea by fifty numbers, and in the normal course of events would be promoted two years earlier throughout their careers!

These manifest injustices and administrative ineptness caused

33

much discontent among the officers affected, which was quite evident to the new lieutenants of 1922. The lieutenants of 1917 were now for the most part captains, and would continue to hold that rank for another twelve years. The 1918 class were first lieutenants, also destined to so remain for a decade or more. In a promotion system based solely on seniority, prospects for advancement did not appear overly attractive.

The enlisted strength, as of June 30, 1922, had fallen to less than 20,000. The Fifth and Sixth Regiments at Quantico were so skeletonized that no company could muster as many as 100 men. The wartime buildings of the Post were literally falling apart, requiring interminable maintenance; consequently, the command was less a military than labor organization.

Commanding the Marine Barracks, Quantico, Virginia, as it was then designated, was a colorful, much decorated extrovert, Brigadier General Smedley D. Butler. He was of the type of daring and unconventional commander best personified in history by the late Brevet Major General George A. Custer, and the later General George Patton. Butler had enjoyed unusual opportunities to achieve fame and glory during his early service, and he had made the most of them. He was a dynamic, ruthless officer, with the demagogic ability to rouse wild enthusiasm in the enlisted men of his command, even though for much of the time they hated and feared him. His professional colleagues did not generally share in this unsophisticated enthusiasm. Be that as it may, Smedley Butler did accomplish much for the physical aspects of Quantico, and he did succeed in keeping the Marine Corps in the public eye during the post-war doldrums.

In June, after graduating from the Basic School, my class received their first assignments to regular duty as officers. I drew a platoon in one of the companies of the Sixth Regiment in time to participate in the celebrated Gettysburg maneuver of that year. We barged the troops up the Potomac River to Washington, then hiked through the blazing summer over the narrow macadam roads of Maryland and Pennsylvania to the Gettysburg Battlefield. Here we encamped for several days, engaged in minor tactical exercises, while Smedley Butler entertained the President and Mrs. Harding in the palatial marquee erected for the purpose by the engineers.

Never shy, Butler leading a cheer at a Marine football game. He frequently
entertained notables (such as President & Mrs. Harding at a palatial marquee at
Gettysburg in a re-enactment of a Civil War battle)

On July Fourth, the Marine Brigade reenacted Pickett's charge, attended fully by the news media of the day, such as Pathe and Universal news cameras. The Marine units were re-designated by the Confederate numbers; companies were listed as Confederate battalions, battalions as regiments, regiments as brigades, etc. Marine officers impersonated individual Confederate commanders. Each Marine was issued a casualty ticket indicating where he should fall. As a battalion leader I "survived" the long march across the open fields and the booming of the cannon, only to fall in the final struggle for the Union battery on Cemetery Ridge. It was all very exciting, and we played our parts with what I thought was professional zeal, before a very large audience, which had come to see us perform. This historical pageant so impressed me that I was able to recall the details some fifty years later when we took our grandson, Tyson, to visit the Gettysburg battlefield.

The following day the Marine Brigade repeated the attack in accordance with modern tactics. No banners, no cavalry charges, no clouds of powder smoke, no heroic leaders waving swords; the open fields were crossed by thin lines of skirmishers, or by rushes of small units, scarcely visible from the Union positions on Cemetery Ridge, presenting only fleeting glimpses of dispersed and rapidly moved targets for the "Yankee" gunners. Even so, the casualties would have been heavy against modern defenses; in the actual attack, Pickett suffered a major disaster. Meanwhile torrential rains had largely inundated our campsite, threatening calamity to the troops. As I recall, it took the combined efforts of the senior troop commanders and the medical staff to persuade General Butler to break camp ahead of schedule.

For me, the bucolic charm of the well kept Maryland and Pennsylvania countryside outweighed the physical discomfort attendant on the march. The troops returned to Quantico sunburned, sore-footed and healthy; but the benefits of tactical training were quickly dissipated as the line companies were bled white by police and maintenance details. My platoon shrank three squads, and I lost interest.

Fortunately, I was shortly reassigned to the Post Rifle Range Detachment as officer in charge of the Pistol Range. My gunnery sergeant was the noted Mickey Finn, who remarked sadly, after observing my first efforts at the targets, "Lootenant, I'll learn ye to shoot that pistol if it takes all summer." He "learnt" me, and it took all summer. The commanding officer of the Range Detachment was Captain Emmett Skinner, whose philosophy was to let his lieutenants "kill their own snakes." I enjoyed the next few months to the fullest.

Later that year, someone in authority noted my record of motor transport experience. I was again reassigned, as assistant Post transportation officer; later to have full charge, under the Post Quartermaster. This was a somewhat exalted position for a new second lieutenant, but fortunately not beyond my technical qualifications. In the discharge of my duties, I came in personal contact with most of the senior officers of the Post, including General Butler. I made many friends, and inevitably aroused the ire of a few who didn't approve of my method of allotting and controlling the available transport. Fortunately for me, my efforts pleased the majority, again including General Butler, and caused me to become more widely known among my seniors than would otherwise have been the case. I believe this fortuitous circumstance favorably affected my later career.

The Marine Corps was then still using motor vehicles of World War I vintage, chiefly Nash Quad, Packard, Mack, and Four-Wheel-Drive trucks, these latter earmarked for "expeditionary duty only." A motley collection of passenger vehicles ranged from Cadillac to Model "T" Ford, including one asthmatic Stutz Bearcat Roadster, reserved as the official mount of the motor transport officer.

The maintenance problem was terrific, particularly during the winter months; and there were frigid mornings when hardly a motor could be made to fire before ten o'clock. An irate and outspoken clientele had to be soothed with such tact as could be mustered, which was good training for youthful Irish temperament.

Came the spring of 1923 and youthful fancies. I had been courting through constant correspondence and periodic visits a young lady whom I had known since college days. She had visited Washington the previous summer, and during my leave period in October we had become formally engaged.

My position in life now seemed to be firmly established. While promotion appeared to be distant, security was assured. Some of my classmates had already married and were well established in Quantico's junior social circles. The life of a bachelor officer did not particularly appeal to me; I felt the need of a stabilizing influence. Accordingly, on June 9, 1923, Miss Nell Neimeyer of Chandler, Oklahoma, became Mrs. Vernon Edgar Megee, of Quantico, Virginia.

Of all the major decisions I have had to make during my career, this turned out to be the soundest. We developed together, absorbing customs and traditions of the Marine Corps to good purpose. Twenty-three years later, the erstwhile lieutenant's bride pinned the silver stars of brigadier on my shoulders, and thus became the general's lady, a position she was to fill with gracious distinction for another thirteen years of active service. Without her by my side, I could not have traveled so far.

Since 1915, the Marine Corps had maintained a brigade in the Republic of Haiti, for the pacification and control of that turbulent country. A similar force had occupied Santo Domingo since 1916. The maintenance of these two major foreign duty stations was the principal concern of the Marine Commandant during the early twenties. All officers and men could expect a two-year assignment there, sooner or later.

My orders arrived for a sailing date in August 1923. Since I had very little "line" experience to my credit, I requested immediate transfer to an infantry unit for refresher training. I had hardly reported to the Fifth Regiment for duty, however, when the orders to Haiti were revoked. We unpacked our meager household belongings and sat down to wait.[*] Again orders were received, this time for an October sailing date. Meanwhile, Smedley Butler had planned another of his epic maneuvers, this time to the Shenandoah to reenact the Civil War valley campaigns. I was promptly tagged as motor transport maintenance officer for the Force trains, my refresher training in infantry tactics sacrificed unceremoniously on the altar of military expediency. My somewhat bewildered bride required little persuasion to go home for a visit.

The "Valley Campaign" was a nightmare of mud for the first ten days. Butler, at this time, was at odds with the city fathers of Fredericksburg, Virginia, and refused to take his troops through that city. Rather, he chose to detour through the red clay hills and pine thickets west of Quantico, over roads unimproved since the Civil War. The September rains caught us on the first day out of Quantico; the solid-tired trucks slid off the narrow clay roads into bottomless ditches; and the infantry regiments had to pitch their pup tents, retrace their weary steps for some five miles and help pull their trains out of the mud. The motor transport officers were decidedly unpopular.

This ordeal was repeated at intervals until we had reached Gordonsville and the improved road net. The troops then moved on in good style over the Blue Ridge Mountains, while the harassed motor transport maintenance officer spent a sleepless week reclaiming their mired and abandoned vehicles. After this episode, I could better appreciate the Civil War history of the area. The generals of that period were not stupid or indolent as has been often charged; they were simply immobilized by Virginia clay.

After participation in the opening phases of the Valley maneuvers,

---

[*] Before actually sailing for Haiti we packed and unpacked three times, a rather frustrating experience for the new bride.

I was flown home to meet my embarkation date. I recall lounging around the unimproved airstrip all one morning waiting for the weather to clear over the mountains. The flight, my second, was made in an awkward stick and wire biplane known as a DT, a designation somewhat disquieting. While the mountains were less than 3,000 feet high, and distant a good ten miles, we yet had to execute a wide climbing spiral to gain the necessary altitude. My pilot was the mustachioed Captain William T. Evans, with whom I was to serve in Nicaragua some six years later.

Haiti: 1923 – 1925

One of the officers ordered to Haiti with us, Captain Winans, had bought a new Model "T" Ford touring car, which he could not drive. He offered it to us for the trip to Norfolk. In that era of winding dirt roads the journey required two days, with a stop over in Richmond. This was a pleasant prelude to our sea voyage.

At Norfolk, we boarded the *U.S.S. Henderson*, a comfortable but slow troop transport, well known to at least a generation of Marines. Our enjoyment of the voyage was dampened somewhat by the cabin assignments. My roommate was a distinguished civilian jurist en route to Haiti on some government mission; my bride was paired with some senior officer's elderly mother-in-law.

Our stay in Port au Prince was brief. Brigade Headquarters had intercepted us with orders to proceed on to Cape Haitian, the north coast outpost of the occupation force, headquarters of the attenuated Second Regiment.

Cape Haitian, the second city of Haiti, was the eighteenth-century capital of France's West Indian colonial empire. It had been burned during the black revolt of 1802, reconstructed by the Haitians, partially destroyed by an earthquake in 1842, and again reconstructed. In 1923, it had a population of some 25,000, mostly black or mulatto, with a sprinkling of whites of various nationalities. The town was located on a natural harbor, snuggled in a narrow crescent between the calm waters of the bay and the steep wooded slopes of Le Morgne, the mountainous promontory, which sheltered Cape Haitian on the north and west. To American eyes, the town and its environs were not without charm, still quaintly French colonial in appearance, customs, and atmosphere. The climate was mild, tempered by the Northeastern trade winds, the daily life of the inhabitants tranquil and unhurried.

Cape Haitian was the market center of the North Coast. To it there came at more or less regular intervals one of the smaller coastal freighters of a Dutch Line, whose arrival with new merchandise was eagerly awaited. An occasional sailing ship would put into the harbor for a cargo of dyewood; and once a month a U.S. Navy transport would call to minister to the needs of the Marine garrison. The days of the cruise ships and swarms of tourists were then far in the future. The only touch of modernity was the Marine mail planes, which dropped in on our salt flat airstrip twice a week or so, the pilots of which were considered dauntless devils for venturing across Haiti's mountains in their World War I DeHaviland two-seaters.

The town was supplied with food by a caravan of patient and overloaded donkeys, guided by colorful and good-natured country women who converged on Cape Haitian during the early morning hours.

Our arrival in Cape Haitian was obviously embarrassing to the official welcoming committee, who for some reason had not been informed that there was a Mrs. Megee along. The other married couples were claimed as house guests by various officers of the garrison and their ladies, while we quite literally were left standing on the dock. After a half hour or so of this unintended ostracism, a Ford car drew up and discharged a very flustered and apologetic Captain Lewis, who took us forthwith to his quarters where we rejoined our fellow passengers at luncheon. My prospective company commander, Captain Ross "Smoke" Iams, who was enduring temporary bachelor status, graciously turned over to us his private living quarters until we could find a place of our own. Here we were introduced to mosquito nets, and such exotic breakfasts as alligator pear drenched with ketchup; it seemed that Captain Iams always had such a breakfast, so his Haitian cook assumed that all Americans had similar tastes!

Our first night in Cape Haitian was marked by the presence of double sentries on the streets, curfews for Americans, and considerable martial confusion. Unknowingly, we started to walk up to the little hotel, which catered to Americans, hoping to find acceptable dining facilities. A smart Marine sentry politely stopped us and suggested that we had better return to our quarters. It seemed that an anti-military occupation editor had that date been released from custody, and some rioting by his followers was anticipated. We returned, dinnerless, to our second-floor quarters and watched the proceedings from a balcony overlooking the main street. This was a rather exciting introduction to our first foreign duty station; but nothing much happened beyond the periodic reposting of the Marine sentries.

A classmate of mine, Earnest Linsert, who had completed his Haitian duty prior to our arrival, had told us that he was reserving for us the comfortable second floor apartment where he and his bride had lived. The keys were to be left with the regimental quartermaster for delivery to us. Unfortunately, we could find no one who admitted knowledge of this arrangement, and we later found that the apartment had been turned over to another officer, a close friend of the quartermaster. Naturally we were rather bitter about this, as habitable houses or apartments were very scarce in Cape Haitian. Nevertheless, we finally located a house, which we thought might do, and set about furnishing it very sparsely with such items as we could pick up locally or borrow from the quartermaster department. We acquired an English-speaking cook and maid-of-all-work and began organizing the household. Almost immediately we both came down with attacks of dengue fever, a bone-breaking ague common to the tropics, which makes the patient very ill for a couple of weeks and incapacitates him for another month. We survived, thanks to the inherent kindness of our old maid, Margaret, and despite the sketchy treatment ladled out by the regimental medical section. It was not a pleasant experience.

My reputation as a motor transport officer had preceded me. After an interval during which I was assigned some nine different jobs, concurrently, I found myself an assistant motor transport officer charged with convoy operations to and from Port au Prince and way stations. The duty was not unpleasant, in fact it was enjoyable, since it enabled me to gain an early understanding of the country, the people, the climate, and, above all, the roads. Such roads as existed had been built by the Marines and their attached naval civil engineers, with native labor more or less impressed for the task. These routes were tolerably passable during the dry season, difficult or impassable during the period of torrential rains. Some of my trips across the mountain ranges and the wide valley of the Artobinite River partook of minor adventures.

During my absence from Cape Haitian on these trips Nell was usually taken in by various ladies of the garrison, so that she wouldn't have to stay alone. Among them was Mrs. Anderson, the middle-aged and motherly wife of an elderly captain, and Mrs. Blum, wife of a first lieutenant in the regiment. Mrs. Anderson, particularly, was most helpful in finding us a more desirable place to live, away from the noisome streets of downtown Cape Haitian. The place selected was at the very end of a cross street, nestling against the mountain, a five-room cottage, called Kie in Creole, with separate kitchen sited in a double-terraced high-walled garden. Entry was by a door in the lower wall. The cottage was of whitewashed masonry with iron roof and wooden partitions, three small rooms in line with a half story over the center room, with a long room on the rear, and a terrace in front overlooking the town and the bay. A large concrete cistern served for a water supply. Our solar-heated shower consisted of a barrel on a curtained stand; the majordomo filled the barrel from the cistern each morning. By late afternoon, the water was warm enough for a bath. The kitchen was equipped with a built-in charcoal stove, although we did have a kerosene range for special cooking. Illumination was by kerosene and gasoline lamps. Primitive though our little house was, we grew to love it.

Our staff consisted of Maurice, a personable youngster, exiled from his upper-class Haitian family for some peccadillo, who spoke excellent English, and who served as our majordomo. Philomese, a sprightly little negress of perhaps fifteen years, was our cook. An older Haitian woman was the laundress. We were living in French colonial style!

Headquarters, Marine Corps, eventually took judicial notice of the fact that a second lieutenant of the Second Regiment was being employed otherwise than "on line duty with troops," in violation of recently established orders. My regimental commander, an irascible eccentric of the old school whose tactical concepts dated from the Philippine Insurrection, acceded grudgingly to this bureaucratic interference with his sacred prerogatives and assigned me to the 53rd

Company for duty, with additional duty in charge of motor convoys!

Eventually, however, I joined my company for a two-month tour at the Brigade Training Center, near Port au Prince. My new captain elected to accompany the troops by sea, permitting us to drive cross-country in his comfortable Buick touring car. My wife had been invited to stay with friends, Lieutenant and Mrs. Orin Wheeler, in Port au Prince while I was occupied with troop training, some twenty miles away. Weekends were mostly free, however, so we were able to enjoy some of the social advantages of life in the capital city, a welcome change from the provincial atmosphere of Cape Haitian.

It might be added here that the social life of Marine officers in Haiti was generally restricted to their own circles. Except for the protocol requirements of the senior commanders there was no social intercourse with the native aristocracy, as was later the rule in Nicaragua, for instance.

The two months spent at the Brigade Training Center were professionally rewarding, marred only by the suicide of my company commander, and ex-sergeant major of long service and some record of previous difficulty. His replacement, also an ex-first sergeant of long service, was an entirely different type, an able and vigorous soldier who wore the starred ribbon of the Medal of Honor and who inspired the instant respect of his lieutenants. Of all the captains under whom I served, Ross Winans stands out as the exceptional troop leader. He taught me the fundamentals of soldiering.

The course of instruction which we followed was progressive and well-arranged, including combat problems with live ammunition, and thorough coverage of the theory and practice of minor tactics. Any company completing this course could have taken the field without further training and engaged in successful combat. This type of training should have been standard procedure for all Marines, instead of deserving special mention here as a unique innovation within the First Marine Brigade. Not until after the beginning of World War II did such individual and small unit combat training become generally prevalent throughout the Corps.

Unfortunately, the results of this carefully organized instruction were not fully realized, since the companies returned to their old garrison routine and were shortly drained of their experienced personnel. Still, the residual effect was not inconsiderable; the junior officers and non-commissioned officers absorbed valuable and lasting lessons in leadership and techniques.

My two-year probationary period was expiring, so I was again ordered to Port au Prince for my competitive examinations. This required some two weeks of daily sessions of intensive cramming. In the event, I was successful, advancing my position on the permanent lineal list by six numbers and my probable date of promotion by three months. I noted with some chagrin that my lowest mark was in "Theory of Marksmanship," a subject in which I considered myself an expert; and that my highest mark was in the subject of "Naval Ordnance and Gunnery," with which I was totally inexperienced. This observation may or may not be construed as a significant commentary on the validity of written examinations.

During the summer of 1924, we were given the opportunity to visit the neighboring Dominican Republic, on the occasion of the withdrawal of the Marine occupational forces. The long day's motor journey, over indifferent roads, was nevertheless refreshing and gave us some appreciation of the countryside and the contrasting Spanish language and culture. The country was completely peaceable; we traveled without military escort and found the people friendly and helpful to travelers.

In October of that year, we availed ourselves of our annual leave and returned to the States aboard the notorious *Kittery*, an ex-German inter-island steamer never designed to be taken out of sheltered waters. The passage of the Virginia Capes was an ordeal, which few passengers care to repeat. We had no choice in the matter, since this was the only naval transport regularly serving the Caribbean area. We spent the holidays in Oklahoma after which I returned alone to Cape Haitian. My wife, Nell, remained in Oklahoma for some three months for further recuperation of her health.

My second year in Cape Haitian was spent largely as the officer-in-charge of the rifle range, with collateral duties in one of the infantry companies. My working day began at six in the morning and ending at noon, except for attendance at the semi-weekly sunset parades and a periodical tour as officer of the day. There was also an occasional escort of celebrities to Christophe's Citadel, a frowning bastion visible twenty miles away across the Plain du Nord.

Our social life was simple and generally enjoyable. For the junior officers there were picnics, boating parties, horseback jaunts over the mountain trails, swimming, exploratory motoring in the old Model "T" touring car, and wild guinea shooting. We remember especially a horseback ride that we made on Christmas Eve, 1923, across the wooded peninsula which shelters Cape Haitian on the north to a long stretch of secluded and picturesque sandy beach, which on this day we had entirely to ourselves. The feeling of utter solitude in this foreign land, was tempered by some uneasiness as to our safety, but added to our spirit of youthful adventure. Nell suffered a moment of terror when the little stallion she was riding decided to enjoy a wallow in the enticing sand, saddle and all. I'm afraid that my bride didn't exactly join in my amusement.

Among our junior colleagues at Cape Haitian were the Ralph DeWitts and the Raymond Coffmans, who continued in later years to be among our most cherished friends. There is nothing like sharing adversity and adventure to seal lasting friendships. Almost forgotten through the softening veil of the years are the tribulations: the bone breaking dengue fever, the primitive appointments of our living quarters, and the isolated monotony of tropical existence. Remembered instead, with an aura of youthful romance, are the rustling palms, the fragrance of exotic flowers, the brilliance of the tropical moon flooding the waters of the bay, and the unforgettable picture of the old square-rigged ship leaving the harbor under full sail with its leisurely collected cargo of dyewood.

Cape Haitian had its share of historical glamour, which was not lost on us. We strolled through the ruins of Pauline's Palace, built long ago for Napoleon's sister who was wife to the French General who commanded in Northern Haiti before the black rebellion. We clambered over the ancient French forts guarding the harbor entrance, inspected the corroded iron relics of Columbus' wrecked ship still visible on the sand spit across the bay, discovered the jungle-choked ruins of once-elegant French plantations, journeyed to Sans Souci and the formidable citadel of Christophe. Life was young and leisurely.

We sailed for home in November 1925, again aboard the creaking *Kittery*, and arrived finally after the usual storm-tossed passage of the

Virginia Capes at Quantico, from whence we had departed two eventful years before. Here my nemesis caught up with me again. I was assigned as assistant motor transport officer of the Tenth Artillery Regiment, and buckled down to the more exacting routine of stateside duty.

Potomac Interlude – 1926-1927

Our first concern was for living quarters, which this time were not available to us on the Post. After several weeks in one of "Ma Gratz's" noisome and overpriced apartments we bought a lot in the town of Quantico and built our own house – a rather daring venture for a young couple in those days. We built well; our little house is still standing and habitable (1978).

That summer I tried some competitive pistol shooting without notable success, and later accompanied the regiment to Camp Meade, Maryland, for the annual session of artillery target practice. We also indulged in some local touring in our newly purchased "Model T" Ford touring car, traversing the Blue Ridge Mountains into the Shenandoah Valley and back through the Harper's Ferry gap to Washington. That fall we managed a trip to the Sesquicentennial celebration in Philadelphia, then on to New York and up to West Point before turning back. On this trip we were accompanied by Nell's sisters, Maud and Blanche, who had been visiting us in Quantico.

I remember that the duck shooting was good that fall along the Potomac and its tributaries, and that roast canvasback was common on our table. Since most of the hunting was for early morning flights, I invariably arrived home with barely enough time to get to work, leaving the duck plucking to my increasingly rebellious bride. She finally arrived at a workable solution – she shaved the ducks, with my razor!

Otherwise, we were enjoying our snug new home, and spent much time improving our house and garden. I even built with my own hands, over a long weekend, what was meant to serve as a temporary garage. We never got around to replacing it. We had hoped and expected to enjoy our occupancy for at least two years, but this was not to be.

I had a busy winter overhauling the artillery tractors that had been worn out the previous year in the building of General Smedley Butler's famous Quantico stadium. Smedley had gone and there was a new regime in Post Headquarters. Major General Eli Kelly Cole had announced in his quiet way that Quantico henceforth was to be known as a military post rather than as an industrial establishment. There was much to be done to create the new image.

Meanwhile, I had been promoted to first lieutenant, and had been given full responsibility for the regimental motor transport. It was time, however, to give some consideration to my future prospects. Over-specialization in a technical branch did not appear to offer much reward beyond the rank of captain. I had tried repeatedly to secure assignment to a line-infantry company, but my technical training had become an incubus of frustration. Perhaps the field of aviation was worthy of consideration; in any event, it seemed more exciting than dull garrison duty. After no little persuasion, I convinced my wife that a transfer to aviation duty would be desirable. My application was accepted with suspicious alacrity by the head of the aviation section at Headquarters, Marine Corps, who had been hard put to balance his input for pilot training against the prevalent casualty rate for Marine pilots.

The diplomatic boiled over in two widely separated places – Nicaragua, and China. The civil war in Nicaragua first drew the Fifth Regiment out of Quantico. We watched our friends sail down the river, wondering when it would be our turn, guessing that we would not have long to wait. It was only a matter of weeks until the Tenth Regiment was ordered to join Smedley Butler's newly formed Third Brigade on the West Coast, for further transport to China. I elected to go with my regiment, after some earnest discussion regarding the desirability of postponing my entry into the aviation branch.

Since this transfer involved foreign duty of indeterminate length, without provision for dependent wives, we planned for Nell to return to her family home in Oklahoma to await further developments. This planned move was precipitated very shortly after my departure for the West Coast by the sudden death of her mother. Nell was faced with accomplishing in a matter of hours all the domestic arrangements, such as renting the house and selling the car, which normally might have been done over a period of weeks. Somehow the task was done and she caught the evening train out of Washington for the long, sad ride home. As it turned out, we were to be separated for almost a year.

Meanwhile, my regiment was making the long trip across country in a mixed troop train, traversing the southern states in the spring of the big floods (1927). The Mississippi River and its tributaries formed at Memphis a vast shallow sea of water across which we crawled on uncertain rails. West of Dallas was then new country to me, and I enjoyed the panorama of desert plain and arid mountain which form the infinite sweep of the Old West. After a chilly interlude in the unheated barracks of the Marine Base, San Diego, we embarked on April 17, along with two battalions of Marine infantry, in the chartered Dollar liner, *President Grant,* for a comparatively luxurious, sixteen-day, non-stop voyage to Olongapo, in the Philippine Islands.

China – 1927

Subic Bay was sheltered, spectacular, and exotic. We released our luxurious cruise ship to her normal employment, and moved ashore at the Olongapo Naval Station, occupying the old barracks built for the previous expedition of 1910. We attempted some training; I do not recall that we accomplished much during our brief stay.

The *U.S.S .Chaumont* steamed down from Shanghai and took us aboard for the return trip. We skirted the east coast of Formosa, speculating as to the extent of Japanese fortifications and other improvements. Off the mouth of the Yangtse, we encountered a whole flotilla of picturesque Chinese junks, whose villainous crews could well have been piratical – undoubtedly were. As we turned into the Whangpoo River we met two Chinese gunboats, which had just concluded an ineffectual bombardment of the Whangpoo forts – so we were told. This was not a particularly impressive introduction to the Chinese civil war, which we had apparently come to monitor.

As we followed the sinuous channel toward Shanghai, we were constantly returning the salutes of smartly manned gunboats and destroyers bearing the insignia of the Rising Sun. Our local pilot held his course, seemingly indifferent to the fate of the tacking junks which constantly cut across our bows. In the anchorage off the Bund, we discovered a French battleship, British, American, and Japanese cruisers, and flotillas of smaller craft. It was an impressive display of international naval power, formally punctilious, yet narrowly watching each other – a harbinger of trouble to follow.

This, then, was my introduction to the fabulous China Coast, of which generations of sailor men have woven tall and lurid tales. The magnificence of the European buildings fronting on the Bund belied the foul sewer that was the Whangpoo, yet could not block the fetid breath of the ancient city that lay behind the marbled façade. The gabble of the wretched sampan coolies as they netted the refuse from our slop chute was only a discordant note of which the old China hands were hardly aware. All these scenes impressed themselves indelibly on my memory, to be weighed in later years when trying to assess the China we knew.

The *Chaumont* tied up alongside the *Henderson* at the Standard Oil compound, some five miles below the Shanghai Bund; here we spent two weeks waiting for a situation to develop. Meanwhile, the Chinese civil war was raging in the North, and it appeared prudent to show the flag in the Tientsin-Peking area. Accordingly, we reshuffled troops and equipment amidst the usual confusion, loaded the Sixth and Tenth regiments aboard the more commodious *Henderson*, leaving the Fourth Regiment to take care of eventualities in Shanghai – a task that was to occupy them for the next fifteen years.

Our route up the coast was laid well out to sea; all we saw was an occasional distant glimpse of the four-stack cruiser that was loosely escorting us. The *Henderson* dropped hook off Taku Bar on June 25, 1927 (my birthday, thus remembered). The shoreline was flat and barely discernible, broken only by the mouth of the Pei Ho River, up which some forty or fifty miles lay Tientsin, our destination. Towed coal barges, grimy and noisome, were provided for our transport up the river, which was accomplished during the night to avoid the heat. Some of the ancient fecal odors which assailed our nostrils from time to time were so pungent as to awaken Marines from a deep slumber, inciting a running and grossly indelicate commentary on all things Chinese.

The brigade commander with his staff and one company of engineers had preceded us on a cruiser. A camp of sorts had been established on Woodrow Wilson Boulevard, in Tientsin. The locale was dusty, malodorous, and stiflingly hot. The engineers had located the mess halls downwind from the "heads," (latrines) so that shortly a mild form of dysentery swept the camp. We were thoroughly uncomfortable, and no little critical of staff planning.

Unloading heavy equipment from the holds of the barges with but a crude derrick and windlass was an interminable task, which proceeded day and night. As each battery assembled its guns, tractors, trucks, and tanks, these were trundled away to assigned gun parks, and the personnel were free to make themselves comfortable. Not until the last vehicle was on the bund could the motor transport officer and his crew relax. This took perhaps a week; it seemed a month.

The heat that summer, aggravated by stifling dust storms, became almost unbearable for North Americans living in tents. Finally, the brigade quartermaster was able lease suitable buildings and compounds (locally, *godowns*), and by the autumn, the troops were in comfortable quarters, preparing for the winter, against which we had been warned. There followed a period of "spit and polish" prescribed by Smedley Butler for the laudable purpose of keeping the troops occupied, but carried to ridiculous lengths by subordinate commanders who hoped thereby to gain preferential notice. Our 75-mm were so polished that they could have been seen in the field for twenty miles on a bright day. Knowledgeable battery commanders ordered buckets of olive drab paint slung from each gun carriage – just in case.

The duck and snipe shooting was good that fall along the canals and in the rice paddies. Further down the river were to be found wild geese, bustard, and large hares, for those who chose to risk possible encounters with roving bands of Chinese brigands. We minimized the risk by traveling in well-armed parties, so were not molested.

The canals froze solid in November, and the Gobi dust storms did not abate. The cold, though, was constant and dry. Men quickly became accustomed to it, and training went on apace. The Chinese armies marched and counter-marched around Tientsin, but left the city strictly alone. Our only enemy proved to be sheer boredom.

Notwithstanding the precautions taken by General Butler and his staff, the fleshpots of an Oriental city proved too much for some of the Marines. When the transgressions became too flagrant, the impatient Butler arbitrarily suspended the disciplinary powers of his subordinate commanders and formed a star-chamber summary court-martial at brigade headquarters for the trial of enlisted offenders. The occasional indiscreet officer could expect no mercy, either, for the Quaker general. They were dealt with summarily and ruthlessly. In the end, discipline and order were restored.

Tientsin "E" Reconnaissance Car in China Expedition, 1927

The American families resident in Tientsin, particularly the officers of the Fifteenth Infantry Regiment, long in garrison there, ameliorated our loneliness with gracious hospitality. They were few, however, and we were many. Our social life was decidedly meager during those long months in China. The uncertainty of our "temporary" status was irksome. Marine officers can stoically endure family separation when there is fighting to be done, but are quick to complain when subjected to personal privation during peacetime.

There were, however, periods of respite worthy of recollection. One such was the weekend excursion to Peking over the Chinese Eastern Railway. Our colleagues in the Legation Guard, a battalion-sized force of Marines, entertained us royally, guided us on sightseeing tours through the temples and lotus lakes of the Western Hills, and on shopping trips through the quaint bazaars of the ancient city. My impression of Peking, then, is dominated by memories of decadent

51

Oriental magnificence, of splendid temples and walled palaces in the midst of incredible squalor and human suffering. The glitter of foreign society in the sacred capital of the Empress formed a symbolic crust over the decaying heart of China. At the time of my visit, however, I doubt I was so concerned with philosophy.

As the months wore on, some of the more affluent officers sent for their wives, whose arrival added to the social amenities of the garrison, but also caused some envy among those not so favored. My wife had planned to join me for Christmas, but was prevented from doing so by illness. In my concern for her, I explored the possibilities of returning to the United States, considering that any military action in China was quite unlikely. Fortunately, a renewed invitation from Washington to transfer to the aviation branch solved my immediate problem.

I sailed from Taku on a bitter day in early January aboard a British coasting steamer, transferred to the *President Cleveland* at Shanghai, boarded the *Chaumont* in Manila, and arrived in San Francisco in mid-February, a post-operative appendectomy case headed for the Mare Island Naval Hospital. After a delayed recovery, my wife joined me in San Francisco and we journeyed to San Diego, where I would begin my aviation training. A mid-air collision, at the very moment of our arrival, took the lives of four Marine flyers, and somewhat dampened my lady's enthusiasm for aviation duty. I had a renewed task of persuasion on my hands. In the end she accepted the situation, with fatalistic philosophy, and never thereafter wavered in her support.

In retrospect, the time spent on the China Coast acquires the virtue of exotic illusion. China, before the Japanese conquest, was a truly fabulous place, where the student could observe the customs of forty centuries past, only lately beginning to yield to the Industrial Age. In the field of agriculture, particularly, the Chinese peasant was predominant. From a plot of land barely large enough for an American kitchen garden, he extracted a living for his numerous family members by the same methods of hand toil known to Confucius. He endured the exactions of venal tax collectors and the ravages of warlords with the same equanimity as his ancestors endured from the hordes of Genghis Khan. Grinding poverty and semi-starvation were his constant companions; he deserved something better than the Communist regimentation that has since become his lot.

As for the foreign resident, he was likely to find the life idyllic. With a highly favorable rate of exchange, the luxuries of China were in reach of even the mid-management groups. Their households were managed by the world's most unobtrusive, efficient, and loyal servants. Their social needs were amply met by the British penchant for clubs and race courses. In truth they lived in the "days of the Empire" – an Empire that has now forever vanished.

Although I was but a detached observer, rather than a participant in the fullest sense, I consider myself fortunate to have known China as it was; the experience was worth the physical and mental discomforts that it cost. The China Tour is one of the highlights of my forty years of military service.

San Diego – 1928

The Marine Aircraft Squadrons, West Coast, were based on the Naval Air Station, North Island, an extension northward of the Silver Strand, which encloses San Diego Bay. The officers of the command, lacking government quarters, were comfortably housed within the adjacent resort village of Coronado. The Navy and the flying Marines shared North Island with the Army Air Corps, whose Rockwell Field then occupied the western half of the island where they previously had been sole inhabitants. Naval Aviation, having secured a foothold, has long since displaced the Army from the area; even their amphibious cousins, the Marines, no longer dispute with them the congested aerial traffic circles.

Here I reported for flight instruction early in March 1928. My immediate hopes for an aviation career were temporarily dampened but the verdict of the Station flight surgeon that I "lacked the requisite visual acuity" to become a pilot. Recalling that I had been temporarily blinded by ether fumes at the time of my recent appendectomy at sea, I managed to obtain a two-month reprieve and continued with preliminary flight training. This was successfully completed; but upon reexamination I was nevertheless found physically disqualified, relieved from flight status, and ordered to the Marine Corps Base, across the bay.

Keenly disappointed at this reversal, we could but accept it.

Before reporting for duty, however, we availed ourselves of a month's accrued leave, bought a Studebaker roadster, and drove to Oklahoma – in those days a motoring epic – since any semblance of paved road ended at the California border. It took us a week each way to traverse the seemingly illimitable desert plains and arid mountains that blocked our passage. Even in retrospect that journey can hardly qualify as a pleasure trip.

Upon reporting to the Marine Base, at that time denuded of troops by the insatiable demands of the expeditions in China and Nicaragua, I was inevitably assigned as motor transport officer, with collateral duty as officer of the day about twice a week. We moved to San Diego proper as being more convenient, and settled down for what we hoped would be a normal tour of duty. In any event, we enjoyed only a brief respite together.

Nicaragua – 1929-1930

Within six months I was scheduled for foreign duty again – this time to Nicaragua, where we were actually engaged in serious guerrilla warfare with the rebel leader Sandino and his cohorts. Somewhat to my surprise, my new assignment was to be with the Aviation branch – as squadrons' quartermaster. I made the trip alone down the west coast of Mexico aboard the *U.S.S. Nitro*, an ammunition ship. In due course we entered Corinto harbor, from whence I proceeded to Managua over the rickety, picturesque National Railway.

I found my new duties engrossing, but not particularly onerous, and I had ample time to enjoy the rather fabulous hunting that was available around Lake Managua. Some of these sporting ventures were recorded in detail in the contemporary pages of *Field and Stream* and other hunting journals. While not required or expected to fly, I did manage to qualify as an aerial gunner – by practicing on alligators in Lake Managua – and was thereafter permitted to participate in an occasional aerial patrol over the combat area.

Daily patrol flights were made to the hill country during the morning hours, to contact and assist the ground units that were patrolling the jungle trails of the Northern Area. Our tri-colored Fokker airplanes were also kept busy with supply and transport missions to this roadless area.

There were frequent clashes and small skirmishes with Nicaraguan "bandit" forces, and occasionally a Fokker would return with wounded or dead Marines. The year 1929 was, however, rather quiet compared with the previous year and the ones to follow. Most of the officers going to or returning from the combat area were overnight guests of the aviation officers' mess, and thus were kept well informed as to the activities of our trail-slogging brethren. Our young pilots would catch a bandit group in the open now and then, just long enough to drop a small bomb or fire one machine gun burst – but this was becoming more infrequent as Sandino's forces learned to move only under cover of darkness or bad weather.

As a matter of record, the weather in the hill country was generally bad, particularly in the afternoons. No pilot wished to be caught behind the mountain ranges much after mid-day – else he might be forced to spend the night as an unwilling, however welcome, guest at some remote outpost where the amenities of life were hardly compatible with the fleshpots of Managua.

The aircraft then available for this pioneer ground support flying were Curtiss "Falcons," Vought "Corsairs," and Loening amphibians, better known as "ducks." All were stick and wire biplanes with fixed landing gear (except for the amphibians), powered by single air-cooled engines of approximately 450 horsepower. For armament, all carried two .30 caliber machine guns, one fixed for frontal firing by the pilot, the other "free," or swivel-mounted, for use by the observer-gunner. External racks on the lower wings were arranged to carry small bombs – 17-, 30-, and 50-pound sizes, later altered to take equivalent weight in 100-pounders. There was no radio or electronic navigational equipment; the pilots flew by compass, dead reckoning, and by the "seat of their pants." Contact was established with ground patrols by means of coded hand signals (often improvised with relatively white undershirts) and message "pickups," these later being accomplished by trailing a lead "fish" from the airplane during a very low pass over a wire suspended from two poles. The hazards of flight were considerable, added to which was the ever-present risk of collecting stray bullets from hidden and unexpected enemy gunners. We suffered an infrequent accident, always resulting in the complete loss of the aircraft. Our aircraft, however, were sturdy and very reliably maintained; our pilots quickly became accustomed at this bush flying, so that our overall casualty rate was probably little greater than it would have been in stateside training.

My wife joined me in Managua after a few months, along with other venturesome ladies, so that Managua social life began to take on some of the aspects of normal garrison duty. Those amenities were not available, of course, to those officers serving in the combat areas, although later some rotation between the Northern Area and Managua was established to ameliorate the situation.

I had made arrangements with the proprietor of a German boarding house in Managua for the temporary domicile of my wife and myself. On the day before her ship was scheduled to arrive in Corinto, however, I was informed that my arrangements would have to be cancelled due to plumbing repairs. In haste, as a last resort, I had to shift to La Hotel Gloria, whose appointments hardly lived up to its name. Our train from Corinto was late – as usual – and by the time we were delivered to the hostelry, the place was dark. After some door banging and some reproachful remarks in my imperfect Spanish, we were admitted and shown to our apartment. The bed was without mattress, boasting only a woven mat over the springs. The plumbing was down the hall. We insisted on dinner, whereupon a chicken was killed and duly served, with scant trimmings. While eating, we were treated to a small but ominous earthquake. This was indeed a sorry welcome to Managua for my lady, but I must say she took it in her stride. After a few days of this we moved into a pension, shared by two other American couples, and presided over by a buxom, avaricious *senora*, whose alleged husband was absent on "business." Here we remained for a couple of months while our permanent quarters were being readied.

We then moved into a small, newly constructed country home adjacent to the airdrome, and within a short walk of the lakeshore. Here we acquired in the course of time a menagerie of the local fauna, including two pet deer whose antics our guests inevitably found most amusing. Life in the environs of Managua was not unpleasant, nor did we find it unduly monotonous. We made some friends among the local Nicaraguans, and acquired a working knowledge of the language, which was to serve us well in later years.

In June of 1930, Sandino returned from temporary exile in Mexico and Honduras, regrouped his forces, and began an alarming sortie toward the coffee *fincas* in Jinotega Province. Our Marine garrisons in this area had been largely withdrawn and replaced by elements of the native *Guardia Nacional*. Sandino's forces intercepted and surrounded a patrol of these on the upper slopes of Saraguasca Mountain, some miles north of Jinotega, inflicted some casualties and would probably have wiped them out but for the fortuitous appearance of two of our aerial scouts. The aviators lent a hand on signal, and did some bombing before returning to Managua for fuel and reinforcements.

A five-plane division was hastily armed, manned, and dispatched to the rescue. I went along as one of the observer-gunners, though no part of duty, merely in the spirit of adventure. We found the beleaguered ground patrol after an hour's flight over the jungle through rain and clouds. The enemy was so well-concealed among the brush and rocks, after their earlier taste of air attack, that nothing was visible even from a very low altitude except a few horses, two of them dead. We bombed and strafed the area thoroughly, permitting the *Guardia* patrol to extricate itself with its wounded under cover of our fire, and join the relieving column from Jinotega. Sandino, although wounded, escaped that night with his followers – with minor loss, as we later learned.

During the attack I twice heard the crack of passing bullets, but was too busy manning my Lewis machine gun to pay much attention/ Upon return to the base at Managua we found two bullet holes in the airplane, one passing through the fuselage just back of my seat. Neither missed me by more than two feet – but they missed. One or two other planes also showed hits from what was obviously a well-manned machine gun. We recovered from the engine nose shield of one plane the cupro nickel jacket of a .30 caliber M1 bullet, of a type but recently adopted by our armed forces, but not yet issued to us. This lent credence to our belief that Sandino's forces were being supplied arms and ammunition across our Rio Grande border. We were later told that Sandino's force included a pair of German soldiers of fortune who were adept with ant-aircraft guns. Obviously our unseen enemy gunners understood moving targets.

This was the first time in eleven years of service that I had been under fire. Nicaragua was, in fact, the first battleground where Marines had been engaged since World War I (save for a minor skirmish or so in Haiti). Thus participation in any combat action was apt to be officially recognized in Washington. Some months later our own episode was duly noted by the receipt of a letter of commendation from the Secretary of the Navy. During World War II this award was upgraded to a Navy-Marine Corps Medal, normally reserved for acts of personal heroism. At the time of the original award, this was my only combat decoration, won, rather ironically, by accidental participation in a mission where it was not my official business to be.

My participation in this and numerous other flying missions did enhance my status among the aviators. Informal flying lessons given by my pilot friends whetted my appetite for aviation duty, in which I was encouraged by the air group commander, Major Ralph J. Mitchell, and other noted flyers such as "Sandy" Sanderson and Frank Schilt. Our flight surgeon was a genial gentleman of Semitic origin, who was pleased to certify that my previously noted visual imperfection no longer existed – if, in fact, it ever had– and that I was "physically and psychologically qualified for duty involving the actual control of aircraft."

Pensacola – 1931

In September of 1930, having previously sent my wife home through the Panama Canal by Naval transport, I boarded a banana boat at Bluefields for the trip to New Orleans, and eventual assignment to the Naval Air Station, Pensacola, Florida, for formal flight training. After a preliminary indoctrination at Norfolk, Virginia, terminating in the traditional first solo flight, we reported in January, 1931, after a long-delayed leave period, to Pensacola for the metamorphosis to winged status.

My first flight instructor was my long-time friend, Lieutenant (later Lieutenant-General) Albert D. Cooley, a kindred soul from Montana. He did a thorough job, not neglecting the small personal indignities with which student pilots were kept in their humble place. Al later told me that he didn't have to teach me to fly the airplane, but that his principal concern was to break me of the persistent habit of landing two feet high!

After nine months of this harassment by successive flight instructors as I moved through progressive stages of training, and after the normal quota of "ups and downs," I was finally awarded the coveted gold wings of a Naval aviator, and immediately assigned to advanced fighter plane training – a plum reserved for the upper bracket of each class. It had taken me three years, in effect, to win my wings. I appreciated them.

There occurred also during our stay in Pensacola another momentous milestone in our lives. On October 17, 1931, in the midst of my busy flight training schedule, our daughter, LaVerne, arrived to give us further stimulation toward success. After having spent five years out of seven on the kind of foreign service hardly compatible with raising a family, and with some prospect of remaining in the States for a while, it had seemed a propitious time to plan for this in truth blessed event. Thus we have had more than one reason to remember, with special fondness, our tour in Pensacola.

In March of 1932, we motored across country to San Diego for our second tour with the West Coast Squadrons. Since this entry into full aviation status was a definite turning point in my military career, I have chosen to end here what I call the preparatory years, and to designate the next stage as the period of development. I was now almost thirty-two years of age, and had served for ten years as a lieutenant. Henceforth, due to my relative seniority among my aviation contemporaries, I was to enjoy more responsible and professionally more satisfying assignments. Although in the beginning my relative inexperience as a pilot was a small handicap, this was compensated for by my long apprenticeship in general Marine Corps duties. Since aviation was definitely a supporting arm within our Service, it was most advantageous to have a sound appreciation of the requirements of the ground elements of the Corps. This advantage accrued to me in increasing ratio as the years passed, bringing increased rank and responsibility.

CHAPTER IV

DEVELOPMENT

West Coast – 1932-1933

The Marine air group to which I next reported for duty was commanded by Major Ross E. (Rusty) Rowell, a quiet-spoken, dignified gentleman, who had learned to fly in his forties after some twenty years of service with the ground elements of the Marine Corps. Rowell had commanded the first air group in Nicaragua, and had pioneered the ground support tactics first used in that campaign. He also established a "post-graduate" flight-training program for all fledgling Marine aviators – to which of course I was immediately assigned, after being informed I was also the new group adjutant.

After some months of these preliminaries, I was then assigned to command the utility squadron – which comprised the transport, photographic, and training airplanes of the group. This assignment gave me ample opportunity to expand my flight experience in the different types, proving to be a professionally profitable interlude.

The Great Depression of the early thirties, under which the country was then prostrate, brought home to us the advantages of the economic security that the Service offered. Civilian friends and relatives who previously had been prosperous were no longer so. We helped carry the family load, where we could – and even withstood the fifteen per cent salary cut imposed by Roosevelt's "New Deal" without complaining overly much.

The Marine Corps reached a low of some 13,000 men in 1933, and we were indeed spread thinly. Fortunately, the China brigade (less the Fourth Regiment) was returned, and the Nicaraguan adventure was liquidated in 1932, thus relieving the pressures of foreign expeditionary service.

The Marines also came out of Haiti in 1934, terminating the role of International police that we had discharged in the Caribbean area since the Spanish-American War. Aside from a small observation squadron, later established on St. Thomas, Virgin Islands, there were no Marine air units deployed outside the United States between 1934 and 1941. Insofar as the Marine aviators were concerned, foreign duty did not exist between the evacuation of Haiti and the beginning of World War II. We had to find a new role.

Up until about 1930, the Marine Corps Schools at Quantico, Va., were but a pale copy of the Army school, still fighting the campaigns of the Second Division in France.

Some of our more far-sighted officers believed, however, that our traditional role should be with the Navy; and that we should cut loose from Army influence to establish an organization and tactical doctrines based on amphibious warfare. The surcease from constant military expeditions and occupations seemed to provide the opportunity for this drastic change of course.

The schools were accordingly reorganized. Committees were set to work compiling new doctrines to cover "small war" (guerilla) and amphibious operations. Officers of experience in the field were invited to submit comments. I recall sending in my views on small war operations, extracts of which I later found incorporated in the new texts.

In the early summer of 1933, orders were received detaching me to the Marine Corps Schools as a student in the ensuing "Company Officers Course." Since completion of this eleven-month course would excuse me from promotion examinations – which at the time appeared to be at least three years distant – I was pleased to have this opportunity to improve my professional status, though reluctant to face the prospect of very limited flying for the year.

The trip cross-country in a new Oldsmobile, over roads considerably improved since our last eastward journey in 1928, and the opportunity to visit our families in Oklahoma, were enjoyable fringe benefits of the transfer.

Quantico – 1933-1935

Classes for the summer session, in topography and Spanish, convened early in June. After the salubrious West Coast climate we found the Quantico summer most irksome. The field mapping trips into the humid forested hills of the reservation served quickly to put us in excellent physical condition – unpopular though such activity might have been for some of our more sedentary classmates. Since I had enjoyed previous experience in topographical mapping and aerial photography this part of the course gave me little concern. My Nicaraguan service proved helpful, also, in the formal study of Spanish, so that the summer session passed pleasantly enough without too much mental effort on my part.

The regular session convened in September, with a strenuous schedule of subjects that required constant application. Scarcely had we settled down to the academic grind, however, when trouble brewed again in Cuba, where the Platt Amendment still imposed on the United States the task of keeping the peace. Headquarters, Marine Corps, could not man the required expeditionary force without pulling officers from the Schools. For two or three days these drafts from faculty and students continued to disrupt the lectures, until finally only a few of us were left, classes then being suspended.

The residual faculty and student body were then formed into committees to prepare the planned new manual of amphibious warfare. I found myself on the aviation committee, with specific responsibilities for the preparation of certain chapters. The ensuing research, writing, revision, correlation, and final editing was an engrossing task – not unlike the preparation of a doctoral dissertation in the graduate school of a large university. The task lasted until the following June, when the aviators were released to catch up on neglected flying, with orders to report back as members of the new class to be convened in September.

While we had suffered a year's delay in our academic program, the committee members had profited commensurately by their research and writing experience. We entered the new class with marked advantage – having just written the texts!

Meanwhile, in the spring of 1934, the Congress had passed a new promotion bill for the Marine Corps, discarding the hoary system for seniority for the principle of selection in all ranks to include first lieutenant. The first selection board to convene wrought havoc in the ranks of lieutenant-colonel and major, which were well-filled at that time by what was euphemistically called "deadwood." This pruning affected, the second board turned to the captain's list, burdened with over age and overly corpulent incumbents, heritage of the post-war personnel policies previously mentioned. More than fifty vacancies were then provided overnight for promotion to captain, and I was fortunate in finding favor with my own board. My elation was only momentarily dampened by subsequent orders to appear before an examination board – not having in hand the exempting diploma from the Marine Corps Schools. In any event, however, the examinations were not such an ordeal; I was duly certified for the promotion which actually came along a couple of weeks later – the mills of the Personnel Department having ground at their usual deliberate pace.

After the Christmas holidays, 1934-1935, I was assigned to the Senior (Field Officers) course to serve as staff aviator in the preparation and presentation of the annual amphibious problem – which, as I recall, dealt with the recapture of Guam, the thesis assuming prophetically that we would lose that island base in the very opening days of any conflict with Japan. I never attended further sessions of the Junior course, but was graduated with the class in June.

The Circus Squadron – 1935-1936

I returned to active flying with alacrity after my years of academic endeavor, being assigned as executive officer of Marine Fighting Squadron Nine, then based on Brown Field of Quantico, commanded by Major Ford O. (Tex) Rogers, a colorful, unconventional, and easy-going flying veteran of World War I, one of the four remaining Marine aviators who had flown in aerial combat – better known in aviation circles as the "Four Horsemen." The air group commander, Lieutenant Colonel Roy S. (Jiggs) Geiger – also one of this elite group – was a great personal friend of Tex Rogers. This did not prevent him from being critically irascible on occasion because of Rogers' disinclination to enforce discipline. Tex turned this problem over to me – which of course was my responsibility as executive officer. Some time later I

63

learned that the air group commander, not without approbation, had referred to me as "Simon Legree." The young pilots of the squadron gleefully adopted this patronymic, and with mock formality presented me with a long blacksnake whip. I accepted this honor gravely and hung the trophy over my office door as a badge of my authority.

The principal mission of VF9M, as the squadron was then officially designated, appeared to be participation in various air shows and public demonstrations, culminating each summer in the Roman holiday known as the National Air Races, held at Cleveland, Ohio. Our summer training schedule, therefore, was devoted entirely to acrobatic maneuvers, formation flying, division takeoffs and landings, until the squadron flew as one plane regardless of attitude while in the air. During this process there was eliminated any pilot who could not meet the grueling requirements of circus flying. The more experienced pilot, reporting to the squadron, was required to first fly in a subordinate position for some weeks before assuming the actual leadership to which his rank might entitle him. Thus was instilled in the wingmen the requisite confidence in their unit leaders. This was sound doctrine. Notwithstanding the obvious hazards of circus flying we had no serious accidents during my tenure.

The squadron was equipped with the stubby short-winged Boeing (F4B4) a stick and wire biplane with fixed landing gear, a top speed of about 140 nautical miles per hour and a diving speed under 200. This plane landed at 60-80 mph, and had an acrobatic maneuverability since unequalled in fighter plane design. It could be, and habitually was, in our squadron, landed in formation on a sod runway less than 1000 feet long, and would take off in half the distance. The Boeing F4B4 was indeed a pilot's airplane, to be remembered with nostalgia in this later jet age. Of course it would have been too slow and under-gunned for the kind of combat the Marine flyers had to face in later years; but this also proved true of the successor models when pitted against the Japanese "Zero" at Midway.

The National Air Races were scheduled for the Labor Day weekend. Our *piece-de-resistance*, calculated to bring the spectators out of their seats, was a takeoff in tight eighteen-plane formation, a full power climb to 10,000 feet, a screaming dive to grandstand level, a zooming wingover with all planes wingtip to wingtip, a second screaming vertical dive and flat pullout, followed by a squadron landing still in tight formation. After three days of this we were drained of exhilaration, and were more appreciative of the role once played by the old Roman gladiator.

Following the Air Races we turned to our neglected military mission. We scheduled our gunnery and bombing exercises; for this purpose moving the squadron, bag and baggage, wives and children, to the old Marine flying field at Parris Island, S.C. The Post was half-empty; few recruits were being trained at the time. We preempted the unoccupied quarters and barracks, set up temporary housekeeping, and took full advantage of the favorable Carolina weather. At the end of the six-week period, all pilots had demonstrated acceptable proficiency with their weapons, and the squadron had become an effective military unit, rather than an assortment of individual aerial acrobats. The stay at Parris Island was an extended outing for the families, as well as a period of unhampered training for the pilots and mechanics. We always looked forward to those balmy months of Carolina Indian summer.

The next major event for which we had to prepare was the annual spring maneuvers of the Quantico Marine brigade, usually scheduled for Puerto Rico, and Culebra, and St. Thomas areas. The squadrons were flown down and back, via Miami, Cuba, Haiti, and Santo Domingo. The ground crews traveled by troop ship, thus leaving the problems of refueling, minor maintenance, and engine starting to the pilots, whose muscles were ill-attuned to the realities of wrestling gasoline drums and winding up inertial engine starters. The maneuver problem, scheduled in January and February to avoid the inclement weather of Quantico, was built around an amphibious landing on the rocky and waterless island of Culebra, preceded by naval and air bombardment of selected targets, followed tactical maneuvers ashore in which the air arm both supported and opposed the ground elements. The observation squadron of the group operated from an improvised field on Culebra; the dive-bomber squadron was based at San Juan with group headquarters and utility units; while the fighter squadron enjoyed semi-independence on the St. Thomas base. This arrangement greatly facilitated the arrangement of mock combats between the attacking and defending elements of the group. All operations this year were in accordance with our newly promulgated doctrines of amphibious warfare, in preparation for the Pacific showdown with Japan, which we all realized would come sooner or later.

After returning from the Caribbean area our squadron was engaged for several weeks in the development of a system of air-to-air bombing, designed for the attack of massed bomber formations. We operated most of the time from the Naval Air Station, Norfolk, Va., in coordination with the patrol bomber squadrons of the Atlantic Fleet. The experiment proved of doubtful value, although in World War II there were instances of such fighter attacks on bomber formations.

In the summer of 1936, in accordance with "career management" dictum, I was ordered to report for the next course of instruction at the Army Air Corps Tactical School (since re-designated as the Air University). Reluctant as I was to leave the active flying that I so thoroughly enjoyed, I acceded with good grace to this plan for continuing my professional education. We took advantage of accrued leave and travel time to journey to Alabama via New England, Eastern Canada, the Middle West, Oklahoma and the Texas Centennial. The fishing in Maine and Wisconsin was relaxing, but the severe drought which prevailed that year throughout the Middle Western states and the Southwest made travel through those states most unpleasant. My wife and daughter took refuge in Colorado while I proceeded to steaming Maxwell Field by rail.

Army Exposure – 1936-1937

The Army Air Tactical School was considered the professional equivalent of the Senior Course at the Marine Corps Schools. It was then the policy to detail each year three Marine aviators of the grade of senior captain to attend this school in lieu of our own. The purpose of this exchange was not only to broaden the professional background of the selected officers but also to assure the Marine Corps the advantages of personal contacts with officers of the other services.

Since the bulk of the student body were army aviators, whose rank varied from first lieutenant to lieutenant colonel, the course was attuned to their needs. Most of them had been flyers since World War I, but had little experience in other branches of the Army. The first three months of instruction were thus devoted to a general review of the tactics and techniques of ground forces, subjects to which the Marine officers were already adequately grounded – or so they thought. After the Christmas holiday we moved into combined air-ground operations and the tactics pertaining to corps and field armies, thus expanding our horizons appreciably.

Classes were held only during the morning hours; afternoons were alternately devoted to equitation and active flying. The opportunity to fly Army aircraft, ranging from lumbering box kites such as the B-4 Curtiss bomber to the latest low-wing monoplane represented by the Northrup A-17, was welcomed and exploited in full measure. The School authorities encouraged weekend cross-country flights, so that visits were made to installations as far afield as San Antonio (where we marveled at the extent of the then-new Randolph Field), to our hometown of Tulsa, and to our home base of Quantico for a brief renewal of the amphibious faith. The Air Corps operation officers blithely assumed that any rated pilot was perforce qualified to fly any type of aircraft – a drastic departure from our own system of rigid checkouts in unfamiliar equipment. I came to appreciate the Air Corps system as fostering rugged self-reliance on the part of the pilots, even though it might be rougher on the airplanes.

During the year, one of our classmates, Major Harold (Little Hal) George, a noted pursuit pilot, organized a volunteer squadron for the practice of experimental pursuit (fighter) tactics. I joined up with alacrity, and soon learned a few new tricks of formation aerial combat. The planes we used were the P-12 (Army version of the Marine F4B4) and the P-6 (Curtiss-Hawk), both being then rated as second-line aircraft, but with which we were thoroughly familiar.

We also flew with a provisional ground-attack squadron that never seemed to get above the 200-foot level during formation maneuvers. For this legal "flathatting" we used the A-12 and the latest A-17, both low-wing monoplane designs.. The Army Air Corps at that period was developing a sound doctrine for the close support of ground units, and we were able to absorb a considerable portion of the tactics and techniques used by their so-called "Attack Aviation" for later use in the Marine Corps.

Oddly enough, when war came, the Air Corps showed little

interest in supporting ground forces during the early operations in North Africa. Only when the Army Ground Forces command registered vigorous complaints at their neglect did the Air Forces reluctantly organize some ground-attack squadrons.

Even then, their system of centralized control of operations rarely gave the ground units the instant cooperation they should have had. The Air Forces were practically autonomous during the War, having expanded so fast that the leadership passed into younger hands, relegating the old timers with ground support experience to secondary roles. By and large these very young generals maintained that air operations were by nature independent of the ground armies, and should be largely confined to the "wild blue yonder." Only during the last year in France did the Air Corps effectively support such freewheeling ground commanders as General George Patton. In the Pacific War, most of the ground support tasks were taken over by the Naval and Marine squadrons.

Nevertheless, In the Marine Corps there were those of us who had profited from the experiments of the old exponents of "Attack Aviation." We carried on their pioneer efforts, tailored to our limited framework of amphibious operations, and by the end of the war had developed a system of air support of ground troops that was the pride of our Corps and the envy of our Army colleagues.

While in later years I frequently had occasion to disagree publicly with the U.S. Air Force doctrines, the year I spent as a guest student at their pre-war school gained for me a wide circle of friends, many of whom became famous air leaders during the War and years later. These contacts were particularly helpful to me during the later part of my active career, and the personal friendships that resulted are still cherished. In retrospect, the year at Maxwell Field was socially most pleasant for us, and professionally most rewarding. We salute the memories of those boon companions of yesteryear – debonair knights of the cockpit.

Academic – 1937-1939

My promotion to the rank of Major was announced while I was at Maxwell Field, although the requisite vacancy proved to be more than a year away. Headquarters, Marine Corps, also informed me that I would attend the next class of the Army Command and General Staff School at Fort Leavenworth, Kansas. Before the event, however, my advance orders were changed. The Marine Corps Schools had magnanimously agreed to take a second aviator on the faculty, and I was "it" – protests to the contrary notwithstanding. So, following graduation exercises in June 1937, our family of three in a two-car caravan drove up to Quantico, where I assumed my professorial role with reluctance and misgiving. Up to that time, the onus of being a "school teacher" was widely believed to be a detriment to continued advancement. In the end, however, the assignment proved to be a "favorable wind" in my case.

After the initial lectures had been laboriously prepared and delivered, I discovered an unexpected flair for public speaking. What I had dreaded now became a pleasure, and I was able to devote my full attention to more imaginative methods of imparting knowledge. Many of my students were kind enough, afterward, to say that they had enjoyed my efforts (I can only hope that some of them may also have profited), so that I was able to feel that I had succeeded in a small way to make aviation popular as an integral part of the Marine Corps.

During the first year I was also charged with writing a new textbook on "The Tactical Employment of Marine Corps Aviation." – an arduous task, which unexpectedly brought me an official "Letter of Commendation" from the commandant of the Schools, Brigadier General James T. Buttrick.

The second year of my professorial tenure was initiated by my actual promotion to the rank of major – a quantum step in those days. The workload seemed easier after that, although I doubt that it was – possibly I had just hit my stride. I was enjoying my work, but felt that it was time for another change. Upon my request, based on the indisputable evidence that I had spent five out of six years on school assignments, I was relieved in June and ordered once more to the West Coast. In taking leave of the Marine Corps Schools, for the last time, I could but reflect on my good fortune, I had been brought into close personal contact with many of the senior officers of the Corps, both on the faculty and within the student body, many of whom were to reach much higher rank in the immediate years ahead (Clifton B. Cates, for instance, would later be Commandant of the Marine Corps). Not a few of these officers would have occasion to remember me kindly at critical points of my own career. Among the more junior students were those who would command battalions and regiments in the battles to come, and whose appreciation of the terrific power of supporting aviation made my own later tasks as an air commander that much easier.

As we were preparing to leave Quantico we participated in one last social occasion of importance. Joyce Geiger, the very comely and sparkling daughter of the then-Colonel Geiger, chose as one of her flower girls our eight-year-old daughter, LaVerne. The wedding was held at the historical Aquia church, located just south of the Quantico military reservation, and was a very formal dress affair. My wife Nell substituted for the bride during the rehearsal, walking demurely down the aisle on the arm of Colonel Geiger. After the actual wedding and during the sprightly reception at the Quantico Officers Club, our daughter, LaVerne, became feverish and had to be taken home. Next morning she was unmistakably afflicted with measles, and was admitted to quarantine in the Posy Hospital. Since our household goods were already packed we were faced with a dilemma. Nell chose to remain with LaVerne at the hospital, while her sister Blanche and I went on to Oklahoma. We were reunited some two weeks later and continued our journey to California.

Carrier Squadron – 1939-1940

Upon reporting to Marine Air Group Two at the Naval Air Station, San Diego, then commanded by Lieutenant Colonel Louis E. Woods, I managed through seniority, previous in fighter planes and some personal persuasion, an assignment as commander of Marine Fighter Squadron Two. This unit then consisted of some twenty-five officers, approximately 100 enlisted men, and twenty active airplanes; these later being the Grumman (F3F3), a stubby biplane with retractable landing gear and an oversize air-cooled engine. Since the fighter squadron of the group had the most modern equipment it was scheduled for later carrier operations with the Pacific Fleet. Naturally, I was elated with my new command and the prospect of an interesting year.

After some weeks of shakedown training, including the indoctrination of new pilots, we began our preliminary carrier qualifications. The initial operations indicated that I had personnel troubles, due chiefly to inexperience in this type flying on the part of one or two of my subordinate flight leaders. I was forced to take rather drastic remedial action in the way of transfers and shifts of flight assignments, in order to restore the morale of the junior pilots whose confidence in their section and division leaders had been impaired by the exhibition of incompetence and bad judgment on repeated occasions. I was responsible, of course, for the performance of my squadron in the eyes of my superiors – who, it must be confessed, had little reason up to that point to be favorably impressed. Added to personnel troubles, we began to suffer material failures: our new airplanes developed a weakness in the landing gear, which would collapse under the strain of arrested landings on the carriers. We suffered several embarrassing deck crashes as a result, before the Group engineering department and the factory representatives could devise a "fix." The crowning misfortune involved the first checkout flight of a new pilot, belatedly assigned to my squadron against my protest. When he did not return to base on schedule, we discovered the fatal crash in the hills back of San Diego, an event not calculated to improve the morale and enthusiasm of the squadron. After this unfortunate incident, the group commander conceded me the veto on any new pilot assignment.

Although the Marine Corps had operated token units from aircraft carriers since 1932, only a comparative few of the older flying officers had enjoyed the opportunity to qualify as carrier pilots. This happened to be my first experience with carrier operations, and I found the flight habits of years rather difficult to adapt to the technique of stalled landing approaches. My age and responsibility as squadron commander did not make the transition any easier – but in the end I succeeded, at the cost of a landing gear or so, bruised ribs, and a somewhat damaged ego.

In late August we took the squadron aboard the old *Saratoga* for our first extended cruise. Our maiden efforts, in comparison with the veteran Navy squadrons aboard, were admittedly ragged, resulting in several minor mishaps and one major landing crash – from which the pilot fortunately escaped without serious injury. The squadron commander habitually landed first and went directly to the bridge to observe the remaining planes come aboard, after which he reported to the carrier captain that his squadron was – or was not – on deck. Each individual landing always had the elements of adventure, if not disaster, so that the ordeal of mentally landing eighteen airplanes after each squadron flight speedily made a fatalist out of any squadron commander who valued his sanity. The carrier captain who shared this concern for *all* the squadrons aboard was understandably calloused and dour, given to caustic comment when performance was not up to par – which it seldom was in the opinion of most skippers.

However, by dint of constant practice, the Marine squadron reached an acceptable level of operational proficiency by the end of the cruise. We had endured considerable good-natured ribbing from our more experienced Naval colleagues, but took our discomfiture in good part. Our day in court would come later.

While the Fleet was lying behind the breakwater in San Pedro harbor, whence it had repaired to give the crews shore leave, we suffered the sudden onslaught of a "Santana" – an offshore windstorm born in deserts of Imperial Valley. Many of the ships' officers and men were on the beach at this time and quite unable to return to their ships. While the *Saratoga* rode out the blow, other ships snapped anchor chains, triggering a wild scramble for the open sea and maneuver room. One destroyer, with only a junior ensign on the bridge, took his ship all the way to San Diego for safe haven – much to the chagrin of his marooned skipper. All in all it was a very spectacular and entirely unrehearsed nautical performance; from my vantage point on the deck of the huge *Saratoga,* I found the Navy not wanting in the emergency,

There was reserved for me, however, one last calamity. As we approached San Diego, all squadrons were spotted for air launch. My squadron was up front; my plane in number one spot. My engine checked out on full throttle test. I nodded to the signal officer and started my takeoff roll down the 200 feet of deck that had been allotted to me. Halfway down I felt the controls firm, lifted the tail preparatory to becoming airborne, expecting only a routine takeoff. Suddenly the engine coughed and stuttered and changed its full-throated roar to a halting staccato bark. It was too late to apply brakes and abort the takeoff; I had no choice but to roll off the bow and attempt to regain partial flight control before striking the water, sixty feet below. The nose of the plane lifted sluggishly, the full right rudder had started a turn away from the carrier's path, when the extended landing gear caught the top of a wave and snagged us down with a tremendous splash. Suffering only from momentary shock and partial immersion I struggled free of the cockpit as the plane bobbed to the surface and floated buoyantly with a steep, nose-down angle. I climbed up on the seat combing as the tremendous hulk of the *Saratoga* slid by me at thirty knots speed, the surging bow wave pushing my derelict plane safely clear of her churning screws.

While waiting for the tailing destroyer to pick me up, I noted also that my squadron had continued launching without loss of interval, joined up smartly on my executive officer, and made a graceful sweep over my ignominious position in farewell salute, en route to the beach and their waiting families. While I had to admire the precision of their drill, I nevertheless felt a twinge of chagrin at the ease with which they could recover from the loss of a squadron commander!

Within five minutes I was plucked out of my briny perch by the destroyer boat crew and hustled aboard for medical examination. The destroyer squadron commander, on whose flagship I was such a precipitate and bedraggled guest, was none other than A.S. (Tip) Merrill, later to become a famous wartime admiral. He apologized for the lack of a proper stimulant in his medical stores, gave me a cup of coffee, a dry shirt, and a seat on his flag bridge for the three-hour trip back to San Diego. Unfortunately, my plane filled and sank during attempted salvage operations, carrying with it a goodly part of my uniform wardrobe. My embarrassment at thus returning "unhorsed" from my first carrier cruise was later ameliorated by receipt of a copy of the carrier division's commander's endorsement on the *Saratoga*'s accident report: "The pilot, Major Megee, handled his plane exceptionally well under circumstances of extreme hazard. s/William F. Halsey." Actually, such handling as there was resulted from instinctive reaction – there was hardly time for planning. Nevertheless, I appreciated the confidence.

73

Into the drink off the *Saratoga*. "You can't imagine how big an aircraft carrier looks as it steams by you at launch speed, with you sitting on a wing."

Vernon, serving as Marine Aide to Admiral Bull Halsey, Hawaii, 1945

In later years Admiral Halsey would introduce me with a chuckle as the "little Marine who got his tail wet off the *Saratoga*. In pre-war days no one ever referred to him as "bull" Halsey; aboard the *Saratoga* he was affectionately known as "Old Sodium Chloride." The *Saratoga* was then commanded by Captain Albert "Putty" Read, who some twenty years earlier had gained fame as commander of the NC-4, first seaplane to effect a trans-Atlantic flight. Halsey's and my trails crossed often in the Pacific during the War, and I was always assured of a warm greeting from one who considered me as an old shipmate. Bill Halsey was in truth an "old sea dog," but with a very human touch which endeared him to all his subordinates. His memory will be revered by seafaring men as long as we have a Navy.

Following the *Saratoga* cruise the carrier squadrons of the Fleet operated for several weeks from shore bases while engaged in gunnery and bombing exercises. These were progressive in nature, beginning with individual pilot qualification and ending with squadron and group tactical exercises employing live ammunition against both stationary and moving targets. At the end of the training period each squadron fired its record practice in competition with all others of the same type, striving to place high in Fleet competition. The culmination of this exciting training period was the bombing by squadrons and air groups of the old battleship *Utah*, which had been stripped of her guns, sandbagged and barricaded, and turned into a very effective and agile target ship. While she couldn't shoot back at us she could still steam at sixteen knots and maneuver violently while under attack. Nevertheless, we managed to bounce a respectable percentage of our 100-pound, non-explosive bombs off her armored decks; and the Marine flyers were not the least skilled in this magnificent sport! During the final phases of the gunnery and bombing exercises I was acting as air group commander for the Marines, so had to surrender squadron leadership to my second-in-command. As it later turned out, this would be my only opportunity to lead an air group in tactical flight.

Most of us realized, that fall of 1939, that we might be engaged in the not-too-distant future in actual bombing attacks against warships flaunting the insignia of the Rising Sun. Our training schedules were devised accordingly, to endure success when the chips would be down. The later demonstrated prowess of the U.S. Naval and Marine aviators in those historic encounters during the Pacific war was not due to accident or natural aptitude; but rather to the months and years of hard and hazardous flying which had developed carrier-borne air operations to such a high degree of efficiency before the fateful bombs were loosed on Pearl Harbor.

Not all our training that year was Naval in character. We found time for cross-country squadron flights – to Salt Lake City, on one occasion – and for ground support training with the Marine brigade stationed across the Bay.

After the New Year there was another short carrier cruise – on the then-new *Yorktown* – in connection with a period of combined air and surface training for the Pacific Fleet. Plagued by rough weather and heaving flight decks, the carrier pilots – mine among them – suffered a series of minor mishaps before "settling into the groove" with the new landing signal officer. I recall that my squadron spent considerable time on simulated air combat patrol, executing lazy circles over the maneuvering fleets twenty thousand feet below. Thus we had in effect a grandstand seat for the study of Naval tactics, which I for one found fascinating.

The final phase of the Fleet problem was a staged air attack against the airfields on the Monterey Peninsula, and as far inland as Reno. The pursuit squadrons of the-then Army Air Corps were the defenders. I recall that my squadron was "jumped" by two Army squadrons of P-35 monoplane fighters while in the process of relieving the combat air patrol over Monterey. We immediately went into our defensive formation of division Lufbery circles, while the Army pilots made the mistake of trying to cut out our slower but more maneuverable planes for individual dogfights. We were armed only with camera guns, of course, but made opportune use of these. When the melee finally broke up among the pine trees and canyons – fortunately without collision – our pilots had a prize collection of close-range "hits" on their exposed film. We later mailed selected prints, suitably enlarged, of course, to the home bases of our opponents. I doubt whether the gifts were appreciated! Years later I did meet an Air Force colonel, James Ferguson, who later became a four-star general, who admitted having been one of the "defeated" squadron commanders in that mock battle, which encounter he said had caused considerable revision in their system of combat training. Thus we learned to survive.

In April of 1940 the squadrons again sought their carrier decks for participation in the annual large-scale Fleet maneuvers, which this year would take us far to the southwest of the Hawaiian Islands. The U.S. Fleet was divided into two major components, one of which was to represent the enemy. Strict radio discipline was enjoined so that neither side could know the position of the other. For several days the carrier squadrons engaged in wide-ranging search exercises, some of which strained our normal radius of action with the fighters, resulting in a close race now and then with the gasoline gauges. Each fighter squadron was accompanied by a navigator in a two-seater observation plane, but the squadron commander was still responsible for the accurate plot of his position while away from his carrier. The use of a plotting board while keeping a relatively unstable airplane in formation required considerable manual dexterity; the resultant "plot" could hardly have been more than roughly approximate. The inevitable bad weather closed in on us one day, and the admirals had to break radio silence to get their airmen home. We were glad to see our rain-swept flight decks again, at whatever cost in disrupted maneuver schedules and irate flag officers. Fortunately, this deviation could not be blamed on the Marines!

Fleet tactics of that pre-war period were built around the battleships, whose ponderous column formed the center of the maneuver circle, round which were posted in deferential disposition first the cruisers, then the outriding destroyers. The carriers, when not launching or recovering aircraft, were required to maintain position astern and on the flanks of the battle line, conforming to the wallowing twelve-knot pace of the dreadnaughts. This restriction on their speed and maneuverability the carrier admirals found irksome indeed, but it took the Japanese submarines to force a change in tactics. In the end it was the new fast battleships that escorted and conformed to the movement of the carrier task forces.

In the weeks that followed, in the balmy South Sea weather, the carrier squadrons had ample opportunity to practice their tactics and techniques. We attacked the opposing carriers, defended our own, escorted our bombers and torpedo squadrons in their attacks against

77

the "enemy" battle line, protected our observation planes while they "spotted" the gunfire of their mother ships – in short, we ran the gamut of Naval aviation tactics. For the Marines, however interesting, this was our *secondary* mission; the Fleet maneuvers of 1940 did not include amphibious exercises.

Finally, both fleets closed in on Pearl Harbor, engaged in a final – and almost disastrous – night battleship engagement, and then put into LaHaina Roads off the island of Maui for rest and replenishment. The Fleet Commander, Admiral Joseph Richardson, took elaborate wartime precautions; destroyers armed with depth charges guarded the anchorage day and night; patrol seaplanes covered the sea approaches for hundreds of miles. This was not part of the maneuvers. The admiral was concerned about the Imperial Japanese Fleet.

The *Yorktown* returned via Seattle and the Bremerton Navy Yard, after some minor exercises en route. Here I had to leave the squadron aboard ship and fly to San Diego to begin preparation for a projected move to an air base on Maui – a plan which was later deferred. I was pleased to be able to report to the Marine air group commander that we had completed the long cruise without even a minor accident; that in fact our spare airplanes were still suspended from the overhead on the hangar deck. Since this was a unique performance, I felt that perhaps for the first time my superior officer was convinced that he had not erred in entrusting to me the command of his only fighter squadron.

Upon the completion of the annual training schedule, which had culminated in the Fleet maneuvers, I had reason to feel that my squadron was ready for whatever duty might come its way. The pilots had been whetted to a keen edge, and the maintenance crews had demonstrated that they were without peers among the carrier squadrons. Our performance had convinced the Naval authorities that Marine Corps squadrons could hold their own at sea, and thus paved the way for their later deployment in the First Carrier Task Forces that finally defeated the Japanese fleet.

My second year of this assignment had scarcely begun when I received orders which were to take me away from tactical flying for three crucial years. I took leave of my squadron with deep regret, little realizing that before I should return to the Fleet Marine Force my old squadron would have formed the nucleus of two new ones, and that many of the pilots and crewmen with whom I had worked and flown would have died on Wake Island or in the initial defense of Midway. Their superb performance against odds, handicapped as they were by inferior equipment, has been historically recorded. I should like to feel that, although denied the opportunity to actually lead them into battle, perhaps some lingering influence of my leadership helped to sustain them during their epic clashes with Japanese "Zeros." I shall always feel a particular pride in their achievements, and an ineffable sadness for those who did not return.

Hunting Andes Deer in Peru with compadres

The end of a successful hunt: Megee is third from the left

CHAPTER V

THE PERUVIAN INTERLUDE

The professional Marine officer may expect, during a normal career, at least one assignment outside the conventional military pattern. However, my receipt of orders to Peru as assistant chief of the newly formed United States Air Mission to that country came as a complete and hardly welcome surprise. I was reluctant to give up my squadron command in that ominous summer of 1940. The reality of having to spend three years away for the Marine Corps proper – even in such an exotic atmosphere as Peru – did not appear to favor my professional prospects.

I also had other reservations about the new assignment. The nominee chief of mission, under whom I would serve on this isolated station, was a senior bachelor colonel with a flair for eccentricity. I had never served under him, and frankly did not relish the prospect; I was equally certain that he had not sought my services. Since he had been selected for this position notwithstanding his ignorance of the Spanish language (for reasons best known to the then-Director of Marine Corps Aviation), it had been necessary that his assistant be able to converse with the Peruvians in their own tongue. Thus my selection for the number two spot. Neither of us had been consulted as to our personal wishes, a not-unusual procedure in the military services, so we could only accept the situation in good faith and go about our preparations.

We went off to Washington together to spend several weeks being indoctrinated by the Office of Naval Intelligence, under whose auspices we were to operate. As we became better acquainted with the new assignment, and with each other, my initial reservations gave way to complete acceptance of what had been . The major part of the indoctrination consisted of intensified exposure to a Spanish language course, conducted by the brothers Saenz, late of Madrid.

Six weeks hardly sufficed to make of my senior colleague a brilliant conversationalist, but it helped. I found the review most helpful since I had scarcely used the language for the ten years past. The brothers Saenz were delightful refugees from the Spanish Civil War, who had a scholarly appreciation of Peruvian history and customs. I recall the parting admonition of Don Roberto: *"Commandante,* you will please to remember that the Spanish you learned on the trails of Nicaragua will not serve you so well in the drawing rooms of Lima." I remembered.

The chief elected to fly the Mission airplane (a single-engine dive bomber) to Peru, while I traveled by sea with my family. We sailed from Los Angeles in early October aboard the *City of San Francisco,* missed our Grace Line connection in Panama, and had to endure a sticky two-week layover in the old Tivoli Hotel. This was tough on my little daughter, who had to surrender her cocker spaniel to quarantine for the period. Eventually, the *Santa Clara* appeared in the Canal, we boarded her in passage from a tender, and in due course, after some eight days of coasting in and out of such exotic tropical ports as Buenaventura, Guayaquil, and Talara, we entered the Peruvian port of Callao, and journey's end – on October 23, 1940.

We found Lima and environs an exciting and fascinating place, and the Peruvian officials most courteous and helpful. After an interlude in the Bolivar Hotel, and the boarding house of Senora Hope Morris, we were installed in a fully staffed suburban villa of considerable pretension, facing the *bosque olivar,* which was to be our home for three full years. Our daughter was entered in a private school to begin her scholastic adventures in the Spanish language. The interminable round of diplomatic calls were made and received, after which we found ourselves caught up in a delightful international society, a stimulating change from the rather parochial social environment in which we had previously functioned.

Some detail of our domestic arrangements might not be amiss at this point in our narrative. Our new house was a two-story Spanish design of white masonry and red-tiled roof, situated within a typical walled garden. The two main wings were joined by a circular hall, off which opened also the entrance to the kitchen hall, the rear patio, and the one-story wing which housed the servants' quarters and the garage. The downstairs part of the main wings enclosed the living room, a small parlor, a library, and a formal dining room. Upstairs were four bedrooms and two baths, reached by a circular stairway rising through the high-ceilinged hall. The architectural detail was tasteful and well arranged for gracious living. In fact, the house and garden were quite the equivalent of any of the several public quarters I was later assigned when serving as a general officer. We were able to afford this luxury due to the favorable exchange rate that made one of our dollars equal to six and one half Peruvian *soles*, the equivalent coin.

The staff was built around Julia, our very able and cultured "Number One Girl," who served in lieu of a *majordomo*, a butler. Since Julia so obviously over-educated for domestic service, we sometimes suspected that she had been "planted" by the Peruvian Intelligence Service, or some other agency. However, she never gave us any reason to doubt her loyalty, and she served us very well indeed. Next in line was Mercedes, a gentle, middle-aged widow who assumed the tasks of personal maid for Nell and governess for LaVerne. She became well loved by all the family, and we maintained some contact with her until her death that occurred in Los Angeles some thirty years later. Mercedes was assisted by the "upstairs girl" and laundress, her sister, Antonia. The kitchen was ruled over by a fourth woman, whose name has escaped us. Adolfo served us as family chauffeur, while a Japanese immigrant took care of the garden even after the Pearl Harbor attack. We found these Peruvian domestics industrious and very loyal to the family; they valued their positions in our household and took pride in their work. In this respect they resembled the European servant class of an earlier time.

Our next-door neighbors were a German family, which with one exception remained strictly aloof. The young daughter of the family, a beautiful, flaxen-haired child about LaVerne's age, made some overtures from the security of her balcony that overlooked our dividing garden wall. LaVerne responded in kind from her balcony, and the interchange became a daily ritual. Neither could speak the other's tongue, but they communicated in Spanish and by gestures, across the barrier wall and across the even more effective barrier of national animosities. We encouraged the friendship between these alien children, but the alienation kept these two little girls from developing a normal relationship. As we came closer to war with Germany, Gretchen was forced to withdraw from the friendship, thereafter affording LaVerne only a fleeting glimpse or a furtive wave. After Pearl Harbor and its diplomatic aftermath, Peru rounded up all German nationals for deportation, among them our neighbors. We had known for some time that the head of our neighbor household was the Peruvian leader of the Nazi party, and we were relieved to have him go; but LaVerne sorely missed flaxen-haired Gretchen and her balcony friendship. We have often wondered what happened to the child; war is so cruel to the innocents.

The Peruvian Air Force, to which we were accredited, was then subordinate to the Minister of Marine and Aviation, as was the Peruvian Navy. The latter had long enjoyed the technical assistance of the United States Naval Mission, whose successful performance had moved the Peruvian authorities to specify that their newly appointed advisory air mission (which was later to replace the Italian Air Mission previously accredited to Peru) should likewise be of the United States Naval Service. Since the Peruvian air arm was predominantly land-based, Marine Corps aviators were logically chosen as advisors.

Shortly after reporting for duty, my Chief was appointed by the President of Peru as *commandante general* of the Peruvian Air Force, as an aftermath of the enforced political retirement of the Peruvian incumbent. While this quite unexpected development gave the U.S. Mission Chief full command authority, rather than a purely advisory function, it could not help but arouse resentment among the senior Peruvian air officers, who thereafter had to deal with their legal commander through an interpreter. Admittedly, this arrangement greatly facilitated badly needed reforms within the Peruvian Air Force. In retrospect, however, it must be doubted that such an innovation in command was basically sound.

In justice to my Chief, it must be said that he discharged his assignment to the full satisfaction of the Minister of Air and Aviation. He even won grudging respect from the rank and file for his diligence in personally inspecting even the most remote of the Peruvian air bases

– and for correcting on the spot the most glaring deficiencies of command and administration. The Peruvian officers of command rank were considerably more restrained in their enthusiasm.

In all these preliminaries I naturally had a part. In all staff conferences I acted as a check on our Peruvian aide and interpreter, and accompanied the Chief on all visits of inspection. Later I was to receive an appointment as deputy chief of the Peruvian Air staff to facilitate our supervision of administrative matters. For the first year, at least, relationships between the Mission Chief and myself were mutually cordial and militarily correct. We were both enjoying the experience of actively directing a foreign military organization, and the sometimes-exciting adventures incident thereto. One such, our first visit to the Amazon frontier bases, could well have been our last. Since this adventure is representative of the flying hazards we then had to face, I shall recount it in considerable detail.

Peru, being south of the equator, has reversed seasons. In mid-summer January the coast is blanketed by a high layer of fog. The high country, on the contrary, is generally clear in mid-winter but subjected to severe storms in the summer months. In the *Montana* region, which comprises the lower eastern slopes of the Andes, and in the *selva* of the Amazon basin, the difference in seasons is measured only by the degree of wetness. January, then, is not the propitious season for flying across the Andes.

Notwithstanding the warnings of the Peruvian aviators, our Chief was determined to have a look at the transmontane air bases without further delay. Our first destination was San Ramon, nestling in the *montana* country directly east of Lima, hardly an hour's flight away. The Chief flew the Mission airplane, which carried water and oxygen equipment. I piloted one of the two accompanying Peruvian Northrop-built monoplanes that afforded neither of these luxuries. In attempting to thread our way through the 16,000-foot pass west of Oroya we became entangled in a snowstorm and lost contact with the flight leader. My Peruvian wingman, Captain Conterno, stuck to me like a leech as we circled and climbed to avoid the encompassing peaks. At 7,000 meters indicated altitude we emerged through a rift in the cloudbanks and caught a brief glimpse of the Mantaro valley, which drains the great interior basin of the Peruvian Andes. Suffering from oxygen deficiency, as evidenced by the black specks dancing before my eyes, I led the way in a downward spiral which brought us over the air base at Huancayo, to ascertain that the Chief was safely down on an emergency airstrip at Jauja, forty miles up the valley.

We later found him in the wretched native hotel, suffering acutely from *soroche*, the mountain sickness. We were directed to return to Huancayo and await his arrival on the morrow. During the night I also suffered an attack of *soroche*, induced by an accumulated oxygen deficiency and the miserably cold rain that drenched the valley. The Chief barely managed to extricate his airplane from the mud of the Jauja airstrip to join us during the late morning. To the amusement of the barrel-chested Peruvians he spent part of the day in the cockpit breathing oxygen. I envied him.

We finally escaped the discomfort and boredom of Huancayo on the third day and found our way through the clouds and gorges to the canyon-locked airstrip at San Ramon, the approaches to which would have tried the flight skill of a condor. The *commandante* of San Ramon, a rather porcine individual, entertained us the best he could for the two days we were marooned on his station. Finally the weeping skies lifted momentarily, barely enough for us to fly down the tortuous canyon to the Ucayli River, which we followed to Pulcallpa for an overnight stay. The airbase at Pulcallpa was even more primitive and lacking in amenities than was San Ramon, although the bachelor *commandante* was more urbane and truly hospitable – to the point of banishing his mistress to the village and turning his house over to us!

From Pulcallpa we followed the river to its junction with the Maranon, where begins the Amazon proper. Even here, some 3,000 miles above its mouth, the Amazon is of truly formidable dimensions. We followed without difficulty the lazy meanderings of the broad ribbon of brown water that eventually led us to Iquitos, the most easterly Peruvian outpost. Colonel Manuel Escalante, of obviously predominant Indian blood, who had received his flight training at Kelly Field in Texas, made us welcome in proficient English. During our two-day stay with him we were made aware of the political intrigue that kept him exiled in such a forgotten corner of the Republic. In fact we soon learned that all the commanders of the isolated transmontane bases were exiles that had been pronounced, for whatever reason, *personae non grata* in Lima. We also discovered that service politics and personal rivalry were not confined to the armed forces of the United States! This intimate disclosure of what went on behind the official façade within the Peruvian Air Force amounted to useful information, which we were later to use to advantage.

Iquitos, in the heyday of the rubber boom, had been an important river port with direct steamer connections to the Atlantic. Now it was but the classic example of a decadent ghost town, boasting a few weathered monuments to its former elegance amidst the squalor and hopelessness of its current status. Prior to the air age, Iquitos was practically inaccessible from Lima except by long sea voyage through the Panama Canal and up the Amazon River. Consequently, the inhabitants of Iquitos found it much easier to travel to European ports than to visit their own capital city. At the time of our first visit (January 1941) periodic commercial air flights were available to and from Chiclayo, on the Pacific coast, connecting there with Lima. This embryonic airline, operated by one "Slim" Fawcett, colorful representative of the earlier barnstorming American birdmen, was not noted for the luxury of its appointments. Flights were made in single-engine Stinsons at no inconsiderable hazard to the passengers. Only those with the most pressing business ventured thus to Iquitos.

On our return journey we ascended the Amazon and Ucayali rivers, then cut across the jungle to Yurimaquas, an upper river port on the Huallaga whose streets had never known a wheeled vehicle. En route, we were careful where possible to observe Escalante's caution to keep our land planes within gliding distance of the rivers or adjacent lagoons. A forced landing in the tangled canopy of the riparian jungle could have had only one result.

At Yurimaguas we were the paying guests of the *Alcade*, who was also the innkeeper and principal trader of the place. While waiting for favorable weather to clear the formidable ranges which barred our way to the Pacific coast we had ample time to observe that Yurimaguas consisted largely of the central plaza, which was ringed by surprisingly substantial masonry buildings in good repair. The lower floors of these were given over to shops, the upper balconied stories serving the owners as residences. From one of these balconies we were greeted by a not-unattractive American woman, who told us that rare indeed were her opportunities to converse with her countrymen. She had lived in this remote Amazonian town for many years; one wondered at the real story behind her exile.

On our way to the marginal airstrip, which adjoined the grass-roofed native *barrio* nestling against the encroaching jungle outside the town proper, we met a number of the aboriginal Indians coming down from the hills to trade. Leading this procession was a Junoesque young woman, magnificently clad in the rosette pelt of a jaguar, who strode regally down the trail looking neither to the left or right nor deigning to notice the strange white men in her path. We respectfully gave way to the queenly presence, who could only have been of very high rank in the tribal high family. In passing we might note that the Spanish Peruvians feared these Indian tribes, whom they had never been able to subjugate, and generally left them strictly alone in their jungle fastnesses.

To reach the coast over the so-called "low pass" of the *cordillera* involved a climb to 5,500 meters and a flight of almost an hour at that level. Again I suffered from lack of oxygen until we encountered the rising heat from the coastal desert, which quickly dissipated the cloud cover and permitted us to descend to a more comfortable altitude. We were happy to avail ourselves of the facilities of the Chiclayo air base for a couple of days, before proceeding without further incident down the coast to Lima.

We later found that the Peruvian aviators had never expected us to return safely. Their experience in the Andean region with their limited equipment and lack of instrument flying techniques had rightly convinced them that ours was a foolhardy and hazardous expedition during that particular season. Perhaps it was, but it served nevertheless to establish us as aviators *formidables*, worthy of their respect. The Peruvian flyers who had accompanied us also gained measurably in stature among their *companeros*.

Although we made several other flights into the Amazon country during the ensuing three years we thereafter went better equipped insofar as aircraft were concerned. None of us again elected to fly one of the Peruvian military airplanes over the mountains. Increased familiarity with the physical aspects of the Peruvian hinterland tended later to reduce the hazards of flying without radio aids to aviation, but we never let ourselves grow complacent about the very real dangers which existed. We all had our share of anxious moments aloft, but no pilot or crewman of the United States Naval Air Mission so much as suffered an injury during our seven-years tenure in Peru. Our successors of the United States Air Force Mission were not so fortunate.

In retrospect, I must consider that the three years I spent in flying the length and breadth of Peru encompassed the most thrilling, the most hazardous, and the most challenging of all my thirty-odd years of flight experience. The journey described was but typical of many others, differing only in detail.

The Peruvian flyers, whose ragtag miscellany of obsolete and castoff military aircraft from Italy, France, England, and the United States, were never designed or equipped to cope with landings and takeoffs from rough airstrips more than two miles high, nor with the violence and unpredictability of the Andean weather, suffered a high casualty rate. They were brave lads, indeed, and skillful, but overly contemptuous of the special hazards attending their flights. We did what we could to impart our knowledge and to improve the dependability of their equipment; eventually they were to receive new aircraft from the United States, which reduced their accident rate.

During the first few weeks of our sojourn in Lima we took the opportunity to make family automobile excursions into the surrounding country; across the Rimac delta and the sand dunes to Ancon, the little seashore resort to the north, where the shifting dunes uncovered and covered again the whitened bones from long-forgotten Indian gravesites; up the Rimac valley past Chosica into the gloomy chasm through which the river, the railroad, and the highway fought for space in their descent from the pass. On one of these latter excursions, which took us on Christmas Day up to the 10,000-foot level, we saw our first train of pack llamas, gaily beribboned and with tinkling bells, as they carefully picked their way down a mountain trail into the road. The ponchoed Indian drivers, who spoke only the Quechua dialect, scarcely responded to our greeting nor noted our rapt appreciation of the exotic spectacle they made. Other excursions followed, south to secluded swimming beaches; or visits to the archeological wonders of Pachacamac. It was all so new and different that we scarcely missed a weekend trip somewhere.

After six months our Mission was augmented by several officers and non-commissioned officers (with their families), so that we could initiate our programs of improving the administration and maintenance of the Peruvian Air Force. Later we were to give some attention to the flight training program and tactical operation of the squadrons. Generally, our efforts were well received, although there was considerable passive resistance from some of the senior Peruvian officers who resented any interference with the established order. The younger and more active flyers were somewhat more enthusiastic about our proposed reforms.

The Peruvian Air Force was then (1941) comparable in size, organization, equipment, and military capability to what Marine Corps Aviation had been twenty years before. The officers and non-commissioned officers were generally intelligent, with at least some degree of technical training. The lower ranks were filled with conscripts from the "cholo" strata of society, largely illiterate and stolid. The rate of pay of all ranks was very low; commanding officers could only maintain their obligations by financial stratagems foreign – and therefore initially unacceptable – to the American military mind. Our progress was frustratingly slow – or so it seemed at the time. Nevertheless, our influence and example bore more fruit as we came to know the Peruvians better, and to appreciate the conditions under which they operated. We developed a genuine liking for each other and formed many close friendships that outlasted the tenure of our Mission.

Toward the end of the first year the Mission Chief took leave for the United States, intending to fly back in a new twin-engine light transport airplane that had been allotted to us. During his absence his duties as chief of mission (but not as *commandante general*) devolved on me. The Peruvian chief of the air staff succeeded to the command. The Mission Chief was absent for some two months, during which the long-smoldering border dispute between Peru and Ecuador flared into open warfare.

The Peruvian Air Force was ordered to the frontier bases in support of the Peruvian Army. The Minister of Marine and Aviation requested that I inspect the tactical dispositions of the air elements and make such recommendations as I deemed prudent. Since our Mission contract provided that in the event of hostilities involving the host country we would be withdrawn, I called the American ambassador (then Henry Norweb) for instructions. He disclaimed any jurisdiction in the matter, so I elected to comply with the Peruvian minister's instructions.

My inspection of the combat area was accomplished by means of flights in the Mission aircraft at a judicious distance behind the Peruvian front lines, operating from the Peruvian airstrip at Tumbes. I tried to be most punctilious in making my official calls on the Peruvian Army and Air Force commanders, and such recommendations as I deemed proper were made informally to the squadron commander based at Tumbes. A pretension of neutrality had to be maintained. Even so, my presence did not go unmarked by the other side. A newspaper in Guayaquil printed a story alleging that Peru was employing foreign aviators – that a Japanese major had been sighted on the air base at Tumbes. My colleagues in Lima pounced delightedly on this piece of information; thenceforth I was to be known as "Major Moto."

The war lasted but a few weeks. The Peruvian Air Force, operating without air opposition, was largely instrumental in bringing the conflict to a swift conclusion. I was sufficiently impressed by this performance to record my impressions for *The Marine Corps Gazette*, a professional military journal, in an article called "An Aerial Blitzkrieg in Miniature," March 1942. This article was picked up, translated, and published in several of the Latin American military journals of the period. The Peruvians, of course`, were pleased at this professional appraisal of their prowess; I had some doubts as to my standing in Equador.

My Chief returned in due course with the new Beechcraft, a twin-engine light transport with the latest radio navigational equipment installed. This machine was to make our subsequent inspection trips more comfortable, and at least a trifle safer. Since the single-engine altitude did not much exceed 10,000 feet, the loss of an engine over the High Sierra country would still have been but a deferred catastrophe. Fortunately, thanks to Messrs. Pratt and Whitney, the engine manufacturers, and to Master Sergeant Fred O'Connor, we never felt any hesitation in going anywhere in the Beechcraft.

It was during this period, just before our entry into the war, that my Chief became involved in a minor diplomatic incident that proved embarrassing to the American Embassy and caused some concern to the officers of the Mission. He had developed a strange aberration about the German Army and the Nazi government and a corresponding trend towards Anglophobia. His public utterances on the subject bewildered the Peruvians and of course enraged the members of the British colony. The British legation registered an official protest. Action at the Washington level was squelched by the Pearl Harbor debacle, which brought an abrupt end to further indiscretions. However, since I had been unable to agree with my superior's expressed views, and had in fact ventured a brief word of caution on the subject, he apparently felt that I had been "less than loyal to my commanding officer." Our relations thereafter became somewhat strained, and for the next year my official life became increasingly unpleasant. I found that my original reservation about serving under this officer was indeed valid.

During the Peruvian-Ecuadorian incident, the United States government had impounded a shipment of military aircraft that the Peruvian government had taken over from Norway, the original contractor. While this move was diplomatically proper for a neutral nation, it aroused fierce resentment among the volatile Peruvians. The popularity rating of the United States missions in Lima dropped alarmingly, with individual members being subjected to icy courtesy, at best. Our Peruvian chauffeur advised us to remove the distinguishing plaque from our car to avoid unpleasant incidents – but this I stubbornly refused to do. Our influence within the Peruvian Air Force was definitely impaired by this incident, since the aviators felt so keenly the loss of their long-anticipated new aircraft. Privately, I could share their disappointment; critically, of course, I could say nothing. The feeling gradually subsided, and shortly we were overwhelmed by events of a graver nature. The Japanese attack on Pearl Harbor brought the Peruvians rallying to our defense – save for the small clique of displaced generals whose sympathies were elsewhere, and who had never ceased their efforts to remove the Mission Chief from his command status.

My first inkling of trouble came in the early afternoon of December 7th. A personal call from the American ambassador requested my immediate attendance at the embassy – a procedure so unusual as to provoke a remark from my wife that "We are probably in the war." As we later listened to the meager and fragmentary details that filtered into Lima during the afternoon and evening, I found the reported situation fantastic and incredible. My previous experience with the U.S. Pacific Fleet when commanded by Admiral Richardson had led me to believe that such a surprise attack would have been impossible. It was of course many weeks before we knew the full magnitude of our naval disaster.

As pointed out to our Chief by the Peruvian general who headed the newly formed independent air ministry, we were now a "defeated nation." He went on to say, with calculated discourtesy, the nuances of which were beyond the limited linguistic ability of my Chief, that it appeared that any nation sending military missions to Peru was bound to become discredited – witness the French, then the Italians, and now the proud Americans. Since my Chief could not answer, and it was not my place to do so, this diplomatic insolence had to go unrebuked. Our prestige as military men suffered acutely during the coming months, and we were able to accomplish but little. Not until after the battle of Midway, which afforded me the opportunity to prepare and deliver a series of lectures to various groups of Peruvian officers, were we able to reassert ourselves.

Immediately after our entry into the War, I had submitted an official request for relief and reassignment to a combat unit. Most of the other Mission officers did likewise. This was disapproved by a flattering return endorsement penned by no less a prose artist that Colonel John W. Thomason, Jr., then head of the Latin American Section of ONI, who pointed out the difficulty of replacing Spanish-speaking officers at such a time. My disappointment was scarcely assuaged by my accelerated promotion to lieutenant colonel, for which I had previously been selected by the 1941 selection board. I realized all too well that interment in such a non-combat assignment would adversely affect my subsequent professional career.

Early in 1942 I was ordered by Washington to return our original Mission aircraft to the United States and pick up as a replacement a new single-engine Grumman amphibian. Since the original aircraft had seen many months of hard service, I anticipated some trouble during the long flight. None materialized, and we made it to Corpus Christi in four days of short hops across Ecuador, Columbia, and Central America. The weather-beaten old dive-bomber that had served us so faithfully on so many hazardous flights had to be surrendered to the Naval Air Training Command and the heavy hands of fledgling pilots. I then proceeded to Washington, via Oklahoma, where I enjoyed a brief family reunion.

I found Washington, those first weeks after the Pearl Harbor disaster, a hectic turmoil; and was glad to escape down the river to our base at Quantico, Virginia, where my new plane awaited me. Quantico was teeming with recalled reservists, with few if any old friends in evidence – they were already on the West Coast bases en route to the war zone. I checked out the new amphibian for what I confidently expected to be a non-eventful flight to Lima, on a rather leisurely itinerary. My first stop was at New Bern, North Carolina, for a look at our yet uncompleted Marine air base at Cherry Point, the magnitude of which I found overwhelming. Next stop was at our newly rejuvenated field at Parris Island, South Carolina, home of a Marine glider group – of all things. (Why the Marine Corps would need gliders was – and still is – beyond my comprehension. We never used them in combat).

So far, my new aircraft was performing smoothly. I had planned my next overnight stop for Pensacola, Florida, via Jacksonville; my early morning departure under a low overcast was without premonition. Ten minutes later a sudden shower of oil darkened my windshield, the oil temperature gauge rose alarmingly and smoke poured from the engine section. Ahead lay a short straight section of a meandering tidal creek, and I did not hesitate on the order of my getting down. I slipped the amphibian in through a maze of fish stakes, felt the drag of the water under the keel, and wound up with a pontoon riding up the slope of a mud bank. The fuselage was covered with oil but fortunately there was no fire. A superficial examination failed to disclose anything wrong with the engine, so after a while we got it running again to send out a radio SOS signal. By this time the ebbing tide had left us stranded in the mud at a rather sharply inclined nose-up attitude. We could only sit there waiting for the air-rescue service from the Savannah air base to find us. Due to the overcast it was almost noon before the first search plane located us, and long after dark before the crash boat could reach us. Meanwhile we lunched on a can of beans and a canteen of stale water while continuing our efforts to localize our engine trouble. Then we were towed sixty miles through a maze of interlocking waterways to a seaplane dock on the Savannah River, arriving at 3:00 a.m. My sergeant-mechanic, one Waldo Harris, who had spent the previous night in ardent wassail with his old cronies at the Parris Island base, slept noisily throughout the towed voyage. For me it had been a long day.

Grumman JF2-5 "Duck" Amphibious Airplane

This was but the beginning of our adventures. After a series of incredible mishaps, which included three more deferred forced landings (in Nicaragua, Columbia, and Ecuador) a week at the Naval Air Station, Coco, Solo, in the Canal Zone, while the experts from Curtiss-Wright completely disassembled the engine without beneficial result, we finally hove into Lima "twenty-four days and four forced landings south of Quantico." It was indeed an epic voyage.

We were later told that our amphibian had a defective oil temperature regulator valve – which we by then well knew – and that several aircraft of this type had been lost on initial ferry flights across the United States. The desk pilots in the Bureau of Aeronautics had followed our movements with marked interest; the betting odds were heavily against our reaching Lima. Cheerful thought! We had already "fixed" the detective valve with a ten-penny nail after our last forced landing at Turbo, Columbia, so declined the invitation to exchange it for a new one. We had no further trouble with the "duck."

The war brought increased military activity to Peru. The military air base at Talara was improved with a new runway and a complete set of new buildings for the Peruvian fighter squadron based there. The International Petroleum Company, whose production and refinery activities were located in that area, undertook the contract. I was the project officer, which required frequent visits to the Talara base and continuing liaison with the Peruvian officials and the American contractors. This gave me the opportunity to become acquainted with some very interesting and colorful fellow countrymen; which contacts might well have helped the Peruvians get a little more than their money's worth. The production manager for International was a lanky, sun-blackened fellow Oklahoman known as "Tip" Moroney, whose knowledge of vernacular Spanish – no less than his technical ability as *jefe* of the drillers – was responsible for the loyalty he inspired among his native employees. "Tip" and I became fast friends, an association which has survived the intervening years.

The floodgates were opened on new equipment for the Peruvian armed forces; funds were made available to us from the United States for the improvement of existing airfields, new equipment for the bases, and – best of all – new military aircraft. In return, the Peruvian government ceded base rights to American air and anti-aircraft units for the protection of the vital refineries in the Talara area. A new American airbase was constructed at Pato Laguna, only a few miles inland from the old Peruvian base, while anti-aircraft units were sited around Talara proper. The resultant influx of American military personnel enlivened the situation for the heretofore-isolated officers of the American missions, and brought home to us some of the realities of our wartime expansion. Units of the United States Navy began to visit Peruvian ports bringing us first-hand information of the war in the Pacific. The American ladies of Lima even organized a thriving USO for the visiting blue jackets and Marines.

The year 1942 was indeed a busy one for the U.S. Air Mission. In addition to my frequent flights to Talara, I did not neglect the other Peruvian bases. My notes of the period indicate that I visited repeatedly the Peruvian airbases and outlying fields in Vitor, Arequipa. Tacna, Juliaca, and Puno – these last two located on or near Lake Titacaca. I also revisited the transmontane bases of San Ramon, Pucallpa, Tingo Maria, Puerto Bermudez, Yurimaguas, and Iquitos. For these trips, which involved long flights over the Amazonian *selvas*, I preferred the amphibian – not only because of the safety factor of being able to land anywhere there was water but also by reason of its powerful super-charged engine which gave it an emergency ceiling of some 25,000 feet.

Notwithstanding this activity, which kept me away from Lima much of the time, and the feeling that I was at last accomplishing something for the Peruvian Air Force, I nevertheless could not escape a sense of futility. Our country was engaged in a desperate war on two fronts, battles were being won and lost, my professional colleagues were distinguishing themselves in combat commands, while I sat out the dance in the service of a non-combatant ally.

In retrospect we were actually accomplishing much more than we realized. We were successfully persuading the Peruvians that in the current war their interests were best served as an ally of the United States, rather than be allied with the Axis powers. When the showdown came Peru definitely and effectively chose our corner of the ring, lending us diplomatic if not actual combat support. I like to think that my three-year effort to make myself *persona grata* to the Peruvian government played some small part in ensuring their loyalty to us. I am also firmly convinced that the Peruvian Air Force would have enjoyed entering combat on our side.

In any event we could only resign ourselves to what had been and make the most of the situation. Our off-duty hours were spent on hunting expeditions to neighboring haciendas for ducks, snipe, yellowlegs, plover, and doves. On one memorable occasion I was invited to hunt *taruga* (Peruvian highland deer) on the vast and alpine reaches of the Hacienda Laive, high above the valley region of Huancayo. We lived in baronial splendor, hunted on horseback attended by a small army of retainers, and indulged in much long-range rifle shooting at the elusive and fleet-footed *taruga*. I brought home three trophies.

Another trip to the highlands was made as a quest of Jeff Morkill, amiable Canadian sportsman and general manager of the Peruvian Central Railway. (Jeff had succeeded his late father in this position). We traveled "over the hill" to Huancayo in Jeff's private railway car, shot ducks and doves in the Montaro valley, tried our hand at fishing for rainbow trout (unsuccessfully), and returned to Lima from Oroya over the pass in Jeff's track inspection vehicle – a Ford station wagon with flanged wheels. The descent from 16,000 feet over the double curves, numerous switchbacks, giddy trestles, and through more than eighty tunnels, was indeed a thrilling adventure, and made me appreciate the engineering genius of the man who could build an operable railroad through such a formidable gorge.

During the late summer of 1942 my Chief was quite perturbed to learn that he had been passed over for temporary promotion to brigadier general, and immediately began heckling his friends in Washington for reassignment to a combat command. This was arranged, finally, which meant that I could have no hope of getting away. Some of my friends in the Embassy suggested to Washington that I succeed to post of chief of the mission, and I had some faint hope that I might. Nothing came of the suggestion, however. Another non-Spanish speaking colonel, who for reasons of his own preferred not to go to the Pacific area, was sent down to Lima in October. This officer was personally much more agreeable – if somewhat less professionally energetic – than was his predecessor, and my personal situation improved accordingly. In November I received my promotion to colonel, which helped me to reconcile to the prospect of another year in Peru.

The Peruvian general who was serving as Air Minister took advantage of this change to take over the *commandancia* of the Air Force by moving into the headquarters offices formerly occupied by the Chief of Mission and myself. This move, while officially unauthorized by the President of Peru, effectively separated my new Chief from his command functions. I objected strenuously to this act of piracy and recommended strongly that the Chief appeal to the President to either uphold his appointment as *commandante general* or cancel it outright, permitting the chief of mission to withdraw with dignity to his original advisory function. This was not done unfortunately; thereafter the titular office of *commandante general* was an empty honor, and the prestige of the U.S. Mission Chief suffered accordingly – as did the work of the mission. Thereafter we could only suggest improvements, whereas formerly we could direct them with authority.

The new Chief had brought down a Lockheed twin-engine transport of advanced design, boasting two-stage supercharged engines for high-altitude work, and the very latest radio navigational equipment. I arranged with the assigned pilot of this craft, a colorful character known as "Blanco" White, to give me a thorough checkout before venturing far away from the Lima area. Thereafter I flew this machine all over Peru; first, to acquaint the new Chief with the Peruvian bases and the country; later, as courtesy pilot for the President, the Minister, and other Peruvian dignitaries. This machine was a great improvement over the Beechcraft, and was a joy to fly. It had a single-engine altitude of better than 16,000 feet, and ample power to take off even from the highest of the Peruvian airstrips – some of which were sited above the 13,000-foot level. What had heretofore been marginal flight operations attended to by considerable hazard now became routine tasks. While flying in Peru with adequate equipment became less of an adventure, I found it no less interesting during my last year. In fact, I was able to develop and report on some new techniques for operating from high altitude fields that were passed on to the Bureau of Aeronautics in Washington and the local officials in Panagra. Instrument and bad weather navigation had become possible with my Lockheed; my personal skills as a pilot were measurably improved. Aside from this satisfaction, however, my last year in Peru can only be described as personally pleasant and professionally frustrating.

During this year we arranged a short family vacation that took us by air to Arequipa, Peru's second city, which nestles against the volcanic cone of Misti at an altitude of 7,600 feet. Here we enjoyed the hospitality of "Tia" Bates, the renowned American lady who had spent her life in the mining camps of Chile, Bolivia, and Peru; and who, now widowed, was the chatelaine of the most famous boarding house in South America. After a few days of adjusting to the altitude – quite necessary for my wife and eleven-year-old daughter, whose activities had been largely confined to the coastal area of Peru – we took the night train to Juliaca and Cuzco, a fascinating ride of almost twenty-four hours. Not until we reached Cuzco did we descend below 12,000 feet; most of the time we were breathing the rarified air of the 15,000-foot level. At Cuzco we took the *auto carril* over the narrow gauge railway across the pass and down the gorge of the Urubamba River to the ancient Inca ruins of Macchu Picchu. This "lost city of the Inca" was to be reached only by foot or on mule back over a tortuous trail which ascended by switchbacks some 2,000 feet above the tiny rail terminus. There were only two mules available to us, so I made the ascent on foot – a feat that proved almost too much for me. The decadent splendor of the "lost city" and the awesome majesty of its natural setting we found quite overwhelming; our visit to Macchu Picchu must always remain as our most outstanding family adventure.

The tour of Peru's southern highlands had brought home to us the exotic customs and colorful costumes of the Quechua tribes, substantially unchanged by the Spanish conquest; the majestic and incredible ruins of the Inca empire; the superposed and tarnished glory of the Spanish colonial era; and the sociological and economic problems of the current republic. The vast and lonely sweep of the bleak *punas*, inhabited only by solitary shepherds and their strange flocks, was somewhat reminiscent of our own arid West. The glaciered peaks and verdant valleys offered the visitor a scenic grandeur not excelled by the best of the Alps or the Canadian Rockies. Yet, in that day, there were few who came to marvel. We have always considered ourselves fortunate to have made the trip toward the end of our tour in Peru – that we might better understand and appreciate what we saw.

Later that year we had planned a weekend motor trip to the resort city of Huaras, located at the base of Huascaran, Peru's highest mountain, a scenic area of Alpine charm. Unfortunately at the last minute I was stricken with a mild dose of dysentery, which had previously spared me notwithstanding multiple exposures. I insisted that Nell and LaVerne go ahead with their plans since they would be accompanying some Swiss friends, the Thommens, long resident in Lima. Our personal chauffeur, Adolfo, who was reliably well versed in mountain driving, managed the journey without incident. The family returned after a week, rejuvenated and enthusiastic about their experience in the Peruvian "Alps." This area has since been devastated by a terrific earthquake and resultant landslide – probably it will never appear the same.

In October we finally received dispatch orders for Washington. In four hectic days we managed our preparations for departure, and made our manners to the *Limenos*. The Minister of Aviation, whom I had not particularly admired, was gracious enough to give me a testimonial luncheon at *Club Union* and present me with the Peruvian *cruz de aviacion* in recognition of my services to his country. Some of my Peruvian friends later congratulated me on my acceptance speech, in Spanish, remarking that I was *emocionado* and *muy gracioso*. Perhaps I was, for I had developed a genuine affection for many of my Peruvian companions of the air trails, which I felt was appreciated.

We elected to fly home via Panagra through Columbia, Panama, and Central America, the submarine menace in the Caribbean and the Atlantic being what it was. Weather conditions altered our normal itinerary so that we enjoyed overnight stays in San Jose, Costa Rica, and Guatemala City, with brief stops in Managua (where we had a fleeting view of our old home on the lake), Tegucigalpa, San Salvador, and Mexico City. Then came the shock of entering, at Brownsville, our own country in the grip of wartime restrictions, ration cards, and travel priorities.

Somehow I managed to get my little family to Oklahoma City, while I went to Washington to learn my new assignment. I was promptly and firmly informed that I was "too old and too senior" for a combat air group command; that I would go to our new base at Cherry Point, N.C., as chief of staff of the Third Marine Air Wing – which I later learned comprised six air groups with some 15,000 officers and men. I had last commanded a squadron of twenty airplanes boasting a total personnel strength of possibly 150! I then began to realize what

101

had been the bemused awakening of Rip Van Winkle.

Before leaving Washington I had occasion to check my official record. To my chagrin, but hardly to my surprise, I found that my former Chief had given me scant credit for my second year's work in Peru. Personal incompatibility had overruled military objectivity, it appeared. At the moment I could but feel regret that I had been unable to fully adjust myself to his eccentricities, and hope that future selection boards would not give undue credence to this one mediocre stain on my professional record. In any event, I need not have worried.

While the Peruvian interlude was a delightful social experience, and the extra compensation received had improved our financial stability, the fact remained that I found myself two years behind my contemporaries in combat command experience. It was now up to me to make up the lost time.

CHAPTER VI

APPROACH TO WAR

Cherry Point – 1943

In mid-November, 1943, the newly completed Marine Air Station at Cherry Point, N.C., with its numerous satellite fields sited within a fifty mile radius, was a roaring, throbbing center of air training activity. The total military population of this vast complex approximated the pre-war strength of the entire Marine Corps (25,000), all dedicated to the urgent task of training squadrons and air groups for effective deployment to the Pacific war zone. We had among our complement pilots and air crewmen in every stage of readiness, from fledglings fresh from the Naval air flight training schools to the seasoned and beribboned veterans of Wake, Midway, and Guadalcanal. Progressive training schedules kept the air filled with airplanes around the clock and the calendar. Into this hive of frenzied activity I was thrust without preparation, a veteran of the diplomatic front.

The commander of the Third Marine Air Wing to whom I now reported was my old friend, "Sheriff" Larkin, now wearing the stars of a brigadier general. He wasted little time on the amenities, as I remember. "Your assignment is chief of staff; your office is across the corridor; Hayne (my predecessor) will show you the ropes. Good luck!"

Thus began a four-month period of intense activity, which in retrospect has certain nightmarish characteristics. My family, who had accompanied me and were installed in one of the new "big houses" on the Station, saw but little of me during the daylight hours. Social activities aboard this wartime station were restricted to the minimum necessary for the military amenities. This Spartan regime proved a sound introduction to the responsibilities and conditions I would later encounter in the war zone. The organizational and administrative problems alone were appalling to one unaccustomed to large-scale operations – not to mention the inexorable pressure of maintaining the training schedule. Somehow, though, I survived this initiation; I even managed after a while to enjoy the heady wine of high command responsibility. I was, in fact, fortunate that I had not been sent directly to the Pacific upon return from Peru.

The Third Air Wing was the senior Marine air tactical command on the East Coast at that time, comprising some six air groups and appropriate support units. As each air group completed its advanced training and was deployed to the West Coast en route to the Pacific area a new unit was organized and the training cycle restarted. A constant stream of individuals was arriving from the basic training schools to merge with the trickle of veterans returning from the war zones to form new squadrons. These had to be located, equipped, and launched into the existing training schedules, later to be incorporated into formally organized groups.

The Wing commander had no responsibility for the command and administration of the Air Base complex; this was the assigned function of the commander, Marine Corps Air Station, Cherry Point – then the redoubtable Frank Schilt, but recently returned from the South Pacific. Each commander reported to a common superior in Washington, thus any disagreement between the tactical commander and the support facilities commander which could not be resolved locally went to the Washington umpire. Such a system of divided command is bound to generate friction; that we got along well at Cherry Point, notwithstanding, is attributable to the personalities and sound common sense of the two commanders and the principal staff officers, and to the camaraderie existing among the older birdmen by reason of their long association in the small, closely knit (pre-war) organization of Marine Aviation. Later, as the war progressed and less-experienced officers succeeded to command, there were instances when the dual command system became unworkable due to the personalities involved. The classic response of higher authority was the summary relief of both commanders. This Draconian penalty probably fostered an injustice here and there; on the other hand the mere threat of such action usually sufficed to maintain compatibility. The dual command system was designed of course to relieve the tactical commanders of all housekeeping chores; since it is still in effect we must admit its validity, with reservations.

One of our most troublesome problems at Cherry Point during this era was the swollen population (some 1600) of lately commissioned reserve second lieutenants from the naval air cadet training program, whose flight training had not gone beyond the basic stage in seaplanes. Furthermore, their only indoctrination into Marine Corps customs, traditions, and regulations had been the issue of a khaki rather than a black tie along with their commissions. Since the Third Wing was geared only to *advanced* tactical training in combat-type aircraft, and had on hand only a very few basic landplane trainers, we could not properly digest these fledglings. They promptly spattered these few airplanes (and all too often themselves) over the countryside, so that in desperation we had to ground them pending action on our urgent request to Washington to have them returned to the Navy for completion of *intermediate* training. Meanwhile, for lack of better quarters, they were billeted in Dallas Huts, six to the unit, which measured 16 feet by 16 feet. As may be imagined the combination of enforced idleness and sordid congestion soon caused the degeneration of their living area into a cross between the Black Hole of Calcutta and a modern big city ghetto. Serious disciplinary problems in due course rose to plague us, to the extent that my morning greeting from my general had a certain Catoesque flavor: "*Something* must be done about these *%#! second lieutenants." Under such pressure the staff *had* to come up with a solution.

A basic officers' military training unit was organized, a very austere camp was established in the most remote corner of the big Cherry Point reservation; staffed by Marine infantry officers of impressive mien and sadistic bent, whose orders were to make Marines of these ex-aviation cadets. As I recall they were given six weeks to accomplish this seemingly hopeless task. Meanwhile the unhappy recipients of this benevolence were restricted to their new pup tent village pending their faintly hoped for transformation into bona fide Marine officers. The final proof was to be an overnight march through forty-odd miles of swampy wilderness to the Marine Base at Camp Lejeune. The "survivors" would then be returned to society with appropriate token of graduation.

The inevitably enterprising reporter visited the camp and wrote his piece for the papers. I was referred to, not too unkindly and unjustly, perhaps, as "Old Iron Pants" by some of the beneficiaries of this Spartan regime – some of whom admitted to me that they actually enjoyed this challenge to their latent military capability. In any event the experiment was a success; the erstwhile cadets became real Marine officers, and later acquitted themselves very well indeed in the air combats of the Pacific War.

Actually, this was but a minor episode in my daily regime. As chief of staff, of course, I had approved the idea, and had inspected the training activities from time to time, bestowing my blessing, so to speak, of the whole idea. I suppose it was but natural that the young officers should have given me the main credit as author of their temporary woes. So be it. I was never averse to accepting any credit that might accrue to my position.

Another and more serious concern of the Wing commander and his staff was the appalling frequency of fatal operational accidents. Young pilots are particularly vulnerable to over confidence and lack of judgment during their period of transition training in combat-type aircraft. Given the urgency of our wartime mission, the often atrocious flying weather prevalent during the winter months along the Carolina coast, and the sheer magnitude of numbers involved, the stage was set for human and materiel wastage. The task of writing condolence letters to the living victims of those weekly tragedies seems more difficult and delicate when death is due to an unglamorous training accident, rather than incident to combat. The patriotic sacrifice is the same, however, in both cases. And in any event, it is a sobering and saddening experience for the military commander to whom such lives have been entrusted.

It was the custom of General Larkin to ceremoniously attend with his staff the departure of a unit for the war zone. Our air groups usually deployed without their aircraft, which were provided within the Pacific area. Thus they embarked as fully equipped Marine infantry troops, filing into their long troop trains after having paid their compliments to the Wing commander in a final ceremony of review. I found these occasions colorful but solemn; the stalwart young Americans marching off unafraid and unperturbed to war, leaving behind the tear-stained symbols of their sacrifice.

As spring 1944 approached, the Third Wing in its turn was ordered to deploy westward, ostensibly to rejoin the tactical groups that had gone before. General Larkin flew out early, leaving the troop movement to his staff – which was normal. After supervising the initial preparations for departure, I turned to the task again of uprooting and resettling my little family. After leaving them in Oklahoma City as an interim residence (they later followed me to the West Coast to await my return from the War), I hurriedly rejoined my command for the cross-country move – which I accomplished by air rather than troop train.

The month of April was spent at the Naval Air Station, San Diego, California, awaiting ship transportation to Pearl Harbor. During this interval I was billeted in the famous old Coronado Hotel, but spent most of my time visiting the various new Marine stations and units, air and ground, which were located up and down the coast and over the mountains in the desert area. This gave me the opportunity of renewing contact with old friends and companions, most of who had already served a tour in the war zone and were willing and able to enlighten me.

The actual voyage to Pearl Harbor was made aboard one of the small converted vessels known as escort carriers, by name the *Hoggatt Bay*. Two accompanying sister ships held the bulk of the wing gear and personnel. The voyage which took about a week was uneventful as far as the enemy was concerned. We were, however, treated to a continuous Roman holiday as the new carrier air group attempted to establish a measure of compatibility between the Grumman "Wildcat" fighter and the grossly inadequate flight deck. The resultant series of deck crashes caused grave doubt in my mind as to the combat capability of such an ill-mated team. A further doubt was raised after I had watched the anti-aircraft gunners of this carrier division attempt to hit a towed target that passed and re-passed them at a comfortable range. To my practiced eye the shots were all passing well astern of the sleeve, indicating crews unversed in the theory of leading a moving target. On this day, at least, an enemy dive-bomber could have sunk us with impunity.

In all fairness, however, it must be remembered that this was my first experience with the wartime Navy, which I perhaps quite unfairly compared with the precision and efficiency of the pre-war fleet carriers on which I had served four years before. I should have made due allowance for the improvised equipment and the hastily assembled and but partially trained crews, whose function in any event was transport – not combat. Later in the war the escort carriers did acquit themselves most creditably in certain types of combat operations.

We entered Pearl Harbor on May 1. Most of the aftermath of December 7, 1941, had been removed, but there were still grim bits of evidence apparent. What impressed me most was the tremendous amount of shipping filling every nook of the vast anchorage. In an earlier day the presence of the entire Pacific Fleet would not have occupied half the space.

I was duly met and escorted to the Marine Air Station of Ewa, to learn that General Larkin had been detached to the South Pacific area for other duty, that the Third Wing would absorb the functions of the defunct command known as Marine Air, Hawaiian Area, and would remain at Ewa permanently as a parent headquarters for transient air groups. Furthermore, I had a new Wing commander, another old friend and associate from Marine Corps School days, Brigadier General "Great" Farrell. After some discussion it was that I would continue to serve as Wing chief of staff.

Ewa – 1944

The summer months of 1944 were spent at Ewa, the Marine Corps air base west of Pearl Harbor. The function of Third Wing Headquarters during this period was that of a personnel clearing and forwarding agency for the forward areas of the Pacific, and as the parent unit of the Marine air groups stationed at Ewa and Midway. I recall making at least one inspection visit to the latter island base – now but a quiet eddy in the torrent of war that had flooded it two years before. I was busy enough with such chores, but was anxious to see some action after my long-delayed entry into the Pacific area.

We witnessed the tremendous buildup of shipping incident to mounting out for the Marianas operation, and followed with enthusiasm the reports of success on Taipan, Tinian, and Guam; being particularly impressed with the news of our significant defeat of Japanese carrier-based aviation by Admiral Spruance's covering Fifth Fleet. We later welcomed the returning heroes of the Marine Amphibious Force staff as they trickled back, freshly bedecked new ribbons and with a convivial desire to share their experiences with the rear-area brethren.

We also witnessed shortly after our arrival at Pearl Harbor probably the most spectacular exposition of pyrotechnics of the entire war, save only for the initial attack on Pearl Harbor by the Japanese Naval Air Force. One peaceful Sunday afternoon, while the congested shipping was loading ammunition for the forthcoming expedition, a mortar shell exploded accidentally, setting off a chain of explosions and fires which eventually destroyed or seriously damaged seven ammunition ships, set fire to the adjacent cane fields, and cost the lives of some 200 men. The holocaust continued for several hours despite the truly heroic actions of the Harbor fire boat crews – an awesome spectacle, the equal of which I did not see in later combat action. It is a tribute to the resourcefulness of the Navy-Marine Corps team that this major disaster was not permitted to interfere with the scheduled sailing of the amphibious armada – although it did seriously deplete the stock of reserve ammunition.

Life in Ewa was comparatively pleasant and comfortable, and there was enough social life to fill the off-duty hours. The Ewa plantation families were most hospitable, as were the few town people we got to know. I recall with special pleasure our relationship with the Doug Bonds, the Slator Millers, and the Bishop Kennedys. Our Third Wing band concert, staged weekly in front of my bungalow quarters, was a popular attraction for our civilian friends – a convenient way for me to repay hospitality. Aside from the social life we had swimming and boating and fishing – to all which pursuits my general paid avid homage. In passing I might add that one of my recurrent chores was finding an aide-de-camp sufficiently skilled in these sports to interest

"Great" Farrell (a former Olympic swimmer) without being so expert as to actually compete with him.

Since Guadalcanal and satellite operations the Marines had been unable to utilize their own air squadrons for close air support of their landings, this function having been assumed by Naval carrier-based squadrons as a secondary mission, performed – somewhat indifferently – under Naval shipboard control. Meanwhile, the land-based Marine air groups were assigned strategic bombing and air defense missions, normally a concern of the U.S. Air Force. After the Mariana operations had again pointed up the inadequacy of the naval air support rendered the Marines, General Holland M. Smith and his colleagues demanded a change.

The obvious defect in the then-prevailing system was the lack of control flexibility. Troop commanders ashore were unable to employ air support to the extent that they could artillery and naval gun-fire, simply because they were not in direct communication with the supporting air units – all requests had to be relayed through naval communication channels to the control center on the AGC (command ship). As a result, practically all air strikes had been of the pre-planned and pre-scheduled type, which although effective in the preliminary stages of a landing, were not responsive to the rapidly changing tactical situation once the battle was joined. The Marine commanders wanted direct control of their supporting air. Our trouble stemmed from the lack of an organization trained and equipped to perform this combat function.

As a first step toward a solution, Marine Corps Headquarters authorized the formation of a new unit to be known initially as the Provisional Air Support Command, charged with the organization, equipment, and training of land-based air control units which could duplicate ashore the functions of the ship-based naval air control units. Probably because of my previous work in the development of air support doctrines, possibly on the recommendation of "Great" Farrell, concurred in by Pat Mulcahy, then CG, Air, Fleet Marine Force, Pacific, I was selected as the first commander of this new unit, due to be established at Ewa early in October.

Meanwhile the Palau operation was in the final preparatory stage, and it was decided that I should join General Geiger's Corps

Headquarters as an air support observer during the Peleliu phase of that operation, an assignment doubly welcome since it would give me some combat experience before I would have to take my prospective command into action early in 1945. Accordingly, in late August, I boarded a NATS plane for Guadalcanal, via Tarawa, to report to III Phibcorps, and then engage in training and rehearsal for landing on Peleliu and Anguar scheduled for 15 September.

After a week or so as a guest at General Geiger's staff mess, during which I had occasion to look over Guadalcanal and adjacent islands, I embarked with the staff aboard the UGC *Mount McKinley* for the final dress rehearsal and voyage to the Palaus. The rehearsal with live ammunition, but only token naval gunfire and air support, was only mildly impressive – mostly because of the obvious errors in execution. General Geiger, however, was kind enough to warn me not to get erroneous ideas about what the real show would be. I had occasion to remember this friendly admonition.

During the five or six days of the approach voyage, I had become familiar with the air support control organization and equipment which the Navy had installed in the new AGC-type command ships. Later I would have to duplicate these facilities ashore. While I found that the Navy had made admirable strides in developing this control mechanism, it did not appear sufficiently flexible from the viewpoint of a marine troop commander. I noted also that the organization included few if any professional naval officers – on the *Mount McKinley* a reserve commander controlled the tactical aircraft for a corps-size amphibious operation. He could hardly be expected to have a comprehensive or sympathetic knowledge of the immediate support requirements of an embattled Marine regimental commander.

The preliminary Naval bombardment of Peleliu was executed during our approach voyage, the apparent results being transmitted to us daily. These optimistic assessments of utter destruction of the Japanese defenses were received with varying degrees of credulity, but, unfortunately gave rise among the troops to the ill-founded belief that Peleliu was going to be "tough but short – three days, maybe two." What a sad awakening these Marines of the famous First Division were to have. In the event, the nut was far harder to crack than even the most experienced and pessimistic officers had feared.

As we approached on the eve of battle tension mounted, the ships were completely darkened and battened down, a condition not conducive to comfort in that tropical climate, aggravated by the usual overcrowded condition. I recall that I had elected to sleep on deck despite the rainsqualls, rather than occupy my allotted cubicle in a steaming cabin. Observers, being strictly supercargo in such a situation, are usually lucky to get a cot on deck! I did enjoy, however, the amenities of the Captain's cabin at meal times, so had no reason to complain of discomfort.

My own feelings on this last evening before D-Day were somber, but not particularly apprehensive. I was then a senior colonel who had served in the Marine Corps for 25 years; and had devoted my life to the study of war without having engaged in any combat action other than one skirmish in Nicaragua. I was thoroughly familiar with the scenario and the script – but had never been on the stage. Tomorrow I would make my belated debut, and that was that. I did not minimize the personal hazard; I simply experienced a fatalistic resignation toward the fortunes of war. I do not recall that these reflections kept me from my allotted sleep. On the morrow there would be work to do.

CHAPTER VII

RECOLLECTIONS OF PELELIU

The great amphibious armada slipped quietly into the allotted transport areas off the southwest reefs of Peleliu, with all hands breakfasted and at battle stations. Lacking a battle station I took a privileged station on the captain's bridge for better observation of the ship-to-shore movement of the small landing craft carrying the combat Marines of the First Division. The prelude, which began at first sign of the tropical dawn, was the awesome bombardment of the landing areas by our supporting squadrons of cruisers and old battleships, which appeared to obliterate the low-lying silhouette of Peleliu with smoke, flame, and dust. The noise from the gun blasts and exploding shells was akin to the continuous rolling thunder of an electrical storm assaulting the Great Plains. After awhile, as the light improved, the landing boats and amphibious tractors could be seen making tight little circles around their mother ships, awaiting their turn at the cargo nets to take aboard the helmeted and heavily equipped Marines.

At the reef line the small control vessels were taking position to affect the transfer of the troops from boats to the amptracs and amphibious trucks (DUKWS). Covering them from the flanks were the destroyers and smaller fire support ships, which now began firing at specific targets on the beaches as the first wave of amptracs approached the reef, hesitated in apparent confusion, then began in a ragged line to clamber over the coral barrier and into the shallow water of the inshore lagoon. Here is where the Marines took their first casualties, not from the Japanese artillery shells that were now dotting the smooth waters with exploding geysers, but from the cleverly concealed mines. I saw one amptrac blown clear of the water, another one or two obviously hit and disabled; after awhile the offshore current was marked here and there by green-clad bodies, floating face down.

The successive waves of amptracs moved inexorably across the shallow water toward the beaches. Now the heavy Naval guns lifted their fire inland to the higher ground along the ventral ridge, and the small rocket ships began their spectacular salvos of whooshing missiles fired over the heads of the advancing Marines. Then with five minutes to go for touchdown the Naval bombardment ceased – to permit the circling squadrons of carrier-based aircraft to come down fore the scheduled last-minute beach strafing. These planes also fired salvos of rockets to the accompaniment of their stuttering .50 caliber machine guns. From my vantage point both Naval gunfire and air strikes seemed effective and according to pre-planned schedules. So far, so good.

Then the first wave of amptracs hit the beaches, only to be immediately enveloped with a hail of accurate and deadly fire from the Japanese anti-boat guns, which had been carefully concealed in fortified bunkers on projecting points and small islets from where they were able to enfilade the landing beaches. Within a minute I could count at least seven burning amptracs at the water's edge. Putting my binoculars on the suspected Japanese gun positions I could clearly distinguish the winking muzzle flashes of the guns. This seemed to me the ideal time to call down the dive-bombers, and I dashed down the ladders to the control room. I might have saved my breath. I was told by the air support commander that he could not divert those bombers without a "request from the troop commander on the spot." Vainly I pointed out that if the troop commander were still alive – which appeared doubtful – he had obviously no means of communication. Fuming at what I considered rank stupidity, I returned to my position on the bridge to watch the continuous slaughter of the amptracs.

Notwithstanding this initial repulse, the waves of amptracs kept coming ashore, hurriedly unloaded their harassed passengers and headed back to the reef for another transfer load. Some didn't make it; presently we could see files of Marines wading toward the beaches. Then, as the morning progressed we could see that the beaches at least were in our hands – although fragmentary reports from the leading units indicated heavy casualties with very little depth to the beachhead. At this time there was desultory Naval gunfire falling on the jungle ridges, and an occasional small air strike was launched against the Jap positions around the airfield. These were hotly countered by the Japanese anti-aircraft crews. I saw one dive-bomber suddenly flame and fail to pull out of his dive; others suffered lesser damage. Few of these carrier planes appeared to be coming low enough in their dives to be really effective, although there were some noteworthy exceptions. In fairness to the pilots, it should be noted they had orders to pull out above 1500 feet. At the moment I could but enter a caustic comment in my notebook.

It was a long, hot afternoon for the thirsty Marines pinned down just back of the beaches. This was especially true of Louis Fuller's First Regiment, which had drawn the left flank position commanded by the fortified Jap positions along the central ridgeline. Herman Hanneken's Seventh Regiment, on the right flank, had somewhat easier going and had made considerable progress by nightfall. Buck Harris and his Fifth Marines were pinned down between the beach and the airfield. Repots from the shore were not encouraging.

All night long the tortured island was kept under illumination by the greenish flare of star shells, fired by the supporting destroyers. Meanwhile our command ship had been moved in closer to the reef but hardly close enough for us to hear the sound of the battle ashore. Behind the Seventh Regiment position, the shore parties struggled to unload and distribute vital supplies and ammunition – not the least of which was potable water. This activity did not go unnoticed by the Japanese; artillery and mortar barrages took their toll here as well.

Aboard the ships we had an air raid alarm, which turned out to be nothing more threatening than a couple of Jap float planes attempting a night reconnaissance on from their base on Babelthaup – a short flight to the north. At the moment, however, it was feared that we might suffer a serious attack from enemy squadrons based in the Philippines. This alarm took a humorous turn aboard our command ship. Ordered by the Admiral to "make smoke" for concealment, the *Mount McKinley* crew touched off the smoke generators located on the fantail of the ship, only to have the equipment flame up, brightly silhouetting the otherwise darkened ship for whatever lurking enemy plane or submarine that might be awaiting such a moment. The resultant megaphone exchange between the Admiral's station and the Captain's bridge relieved an otherwise tense situation and created a wave of audible mirth on the lower decks.

By noon on the second day the beachhead had definitely been established, at a very high price in casualties. Almost every ship's boat returning from the beach brought a freight of wounded Marines, who were taken aboard whatever vessel that offered medical facilities. Reports from ashore indicated heavy attrition from heat exhaustion, as well as enemy action, and demanded a better supply of fresh water. Ships' boat crews volunteered to carry cold water to the front lines, and were duly rewarded with souvenirs – and in a few cases, Purple Hearts.

Since there was little to see from topside at this stage I spent most of the day in the control room, monitoring the air support circuits. About five o'clock we picked up an excited message from a dive-bomber pilot reporting the start of a tank counter attack from behind the ridge north of the airfield. He was cleared for an immediate attack on this column, while the corps operation was notified. The air attack failed to stop the Jap tank column, which crossed the airfield at high speed and slammed into our lines. All was confusion for a few minutes, according to our air observer, after which no moving tanks could be seen. We were to see firsthand, next day, just what had happened.

On the morning of D plus 2 days, I was invited by the corps commander, General Geiger, to accompany him ashore for a front line inspection. As I recall, this first trip was made across the reef in an amptrac, as the Navy had not yet placed the pontoon barges in position for an unloading pier. We visited Brigadier General Oliver P. Smith, the assistant division commander, in his precarious C.P., located in a section of a Japanese tank trap that afforded a modicum of protection against mortar and artillery fire, but no amenities whatever. I recall his pithy remark, as we viewed the scenes of destruction all around us: "Megee, this is a helluva way to make a living!" And so it was.

I found Colonel Herman Hanneken (a medal of honor officer from the early days in Haiti) to apologize for an earlier mishap in which a misdirected aerial rocket had killed a couple of his 7th Regiment

Marines. His answer: "Don't let it worry you, Mac; those planes have saved a great many lives. I wouldn't want to cramp their style." This was the first time I had encountered this appreciative attitude among the ground commanders. Later on it was the rule.

While approaching the airfield I had continually to dodge the amptracs that were streaming across the sand with loads of artillery ammunition. Only yesterday this had been a battlefield, and dead Japs were lying about, some flattened into the sand by the crawler tracks of these lumbering vehicles. What otherwise might have been shocking was accepted as commonplace in this incredible tangle of shell craters and shattered jungle growth. There had simply not been time to police up this section of the battlefield, although I saw here none of our own dead.

At Bucky Harris' command post, which had suffered a direct mortar hit the evening before, I found a somewhat shaken regimental staff. Bucky himself, an old friend from my China days, was suffering mildly from shell shock, but his regiment had crossed the airfield and captured the Jap bunkers on the far side of the ridge. They had paid a price. I walked out on the runway, ostensibly to check for necessary repairs before our own planes could be brought in, actually to see for myself the aftermath of the battle. On the far side I noticed some shapeless bundles of gray-green clothing, scattered here and there like sheaves of wheat behind a grain reaper. I approached one of these to discover the body of a young Marine, lying on his side as though in peaceful sleep. He had died instantly from a bullet wound through his chest. I picked up his rifle and slid back the bolt to find the chamber loaded and ready to fire. Profoundly shocked at this experience I moved down the runway toward the beach, carefully avoiding the remaining bundles.

At the end of the runway, at the edge of the shattered jungle growth, were the burnt-out remains of the Jap tank column that had sortied in counter attack the evening before. Draped across their scorched turrets and scattered around the broken tracks were the mangled bodies of the crews; saturating the humid tropical air was the sickly sweet odor of death. One of our Sherman tanks with a broken tread was parked just within the brush with the turret gun swung to cover the half circle of enemy wreckage. What had happened was clear. Our disabled tank had apparently picked off seven of the lightly armored enemy vehicles as they circled past him at high speed. I was reminded of Remington's picture of the old scout besieged by mounted Indians at the waterhole.

The other ten enemy tanks had been stopped by the Marine infantry, using bazooka, grenades, machine guns, and rifles. The bodies of two Marines were in a shell hole nearby; otherwise no one was in evidence on this eerie scene. The current of battle had flowed by, as evidenced by the crump of mortar shells and the crackling of rifles at the base of the ridge. The burial details had not yet reached the area where I stood – reflecting on the horrors of war as apparent on first view of a battlefield. It then occurred to me that I might make a tempting target for a Jap sniper hidden in the caves of the ridge above. I turned into the screening jungle and rejoined my party for the ride back to the ship.

During the day that I was ashore on Peleliu (Sept. 17) the 81st Army Division executed a lightly opposed landing on the small island of Anguar lying some five miles SW of Peleliu. The following day I again accompanied General Geiger and some of his staff on an inspection of the two Army beachheads. While we could hear sporadic firing from the high ground inland, there was nothing to compare with the concentrated fury of Peleliu. Here the Japs had only a limited force – probably not more than a battalion. As it was, the green Army troops, two regiments strength, took at least a week to secure Anguar. It was hardly worth the effort, considering the amount of Naval gunfire and air support wasted on it – which could better have been used to ease the Marines' burden on Peleliu.

The most prominent landmark on Anguar was a black and white lighthouse perched on the most northerly point. This of course made an excellent reference point for laying down Naval gunfire; some of the supporting destroyer gun crews could not resist using it for an actual target – thereby bringing down on their heads some irascible comment from the task force commander, Rear Admiral George Fort.

Meanwhile, on Peleliu, the Marines continued their dogged and costly advance against the formidable Jap defenses on "Bloody Nose" Ridge. Louis Puller's First Regiment had lost more than 50% casualties and was definitely stopped on the left flank. I was with

General Geiger when he visited Major General Rupertus, who with a broken ankle was trying to command the First Marine Division from a hospital cot. Geiger, as corps commander, insisted on bringing over an Army regiment, until then being held back as corps reserve, to take over from the decimated First Marines. Rupertus demurred stubbornly, holding that such a relief would be a reflection on the Marine Division. Geiger, equally stubborn, but far better informed as to the true situation, nevertheless ordered the relief. A couple of days later I was able to visit Louis Puller and the remnants of his regiment in their "rehabilitation camp" among the coconuts along the SW beaches.

By this time the Fifth Regiment had pushed on well beyond the airfield to the East Coast beaches, and I was able to inspect the wreckage of the Jap hangars and some of their medium bombers that had been caught on the field by our preliminary air strikes and Naval bombardment. I recall passing a number of anti-aircraft gun positions around which were still laying the desiccated bodies of their crews. We had trouble enough burying our own dead on Peleliu; these would have to wait for the cleanup squads of the Island Command troops.

Not until September 24 was the airfield safe for tactical aircraft, other than the small "grasshoppers" of VMO-3, which had been most ably performing their normal chores of tactical reconnaissance and artillery spotting. A contingent of F6F night fighters from VMF (N) 541 were the first to sit down on the hastily cleared and patched runway. They were followed within a day or so by a full squadron of white-nosed Corsairs, commanded by the colorful and able young major from Montana, Robert F. "Cowboy" Stout. These air units were part of the Second Marine Air Wing, then commanded by Major General James T. "Nuts" Moore, my former chief during the Peruvian interlude. While I was not particularly anxious to renew my acquaintance with this eccentric officer, I immediately established close liaison with his operational staff and squadron commanders. As a result of these conferences, which included Colonel Willie Harrison, the Artillery Commander of the Division, no time was lost in putting the Marine aviators to work on their first priority mission – supporting the hard-pressed ground troops.

The chief obstacle was the artillery positions that were south of the airfield, and whose trajectories crossed the airfield at low altitudes.

Stout to take off from under these lethal trails, turn back over them and do his bombing, and then come in for a landing with a very low approach. This system worked like a charm and resulted in the shortest bomb runs in the history of air warfare. The Corsairs never even raised their landing gear after takeoff and were back on the runway within five or ten minutes. The artillery never had to stop firing. The ground Marines were elated at this most effective addition to their fire support, and for the first time during this campaign I put aside my feeling of frustration over the inadequacies of supporting air – as practiced by the Navy.

The technique used by Stout's Marine pilots was most effective. They used 1000-pound bombs with delayed fuses and defused napalm bombs. These they dropped on pinpoint targets from very shallow dives that started under 1500 feet and pulled out at ridge-top level. The ground Marines then fired white phosphorus shells into the spilled napalm liquid, creating very devastating fires at the mouths of the coral caves. The heavy bombs penetrated deeply into the honeycombed coral and opened up the Jap defenses. The entire operation, which continued for several days, enabled the ground Marines to take "Bloody Nose" Ridge and its supporting pinnacles for their own. Since the Marine planes were never out of sight of the troops from takeoff to landing, and their cargo was dumped practically in the laps of their foot-slogging brethren, considerable enthusiasm was generated for Marine Aviation – until then a practically unknown entity to perhaps a majority of the First Marine Division rank and file.

About this time I accompanied the corps chief of staff, then Colonel Merwin Silverthorn, on a tour of the Marine ground positions. We crossed the airfield, entered "Death Valley" at the base of "Bloody Nose" Ridge, then worked our way up a transverse ridge to the battalion CP of Lt. Col. Spencer Berger, whose Marines were leaning against Jap defenses. During this approach we were escorted by a sallow-faced sergeant armed with a Browning automatic rifle which he carried ready at his hip, and whose eyes never left the caves and fissures of the adjacent coral cliffs. At Berger's CP the still-warm body of a Marine lay under a blanket, his bullet-perforated helmet by his side. Berger was wan from strain and lack of sleep, his face covered by a mat of heavy black beard. I told him, with perhaps ill-timed humor, that his CP reminded me of the lair of Jessie James and his Missouri bandits. I recall that Berger told us that he did not expect to survive this particular operation; he was really "down." I reminded him of his premonition when we chanced to meet again some years later, in Naples.

Silver and I then took a jeep up the West Road toward the north end of the coral ridges. The road was strewn with the wreckage of war, burned out vehicles and tanks, shattered Jap anti-tanks guns and the still-unburied enemy dead. The Marines were by now on the ridges, digging their opponents out of caves. We walked up to the crest and carelessly looked over toward some of the enemy positions. There was a sibilant whispering over our heads, too close, as we suddenly realized that the Jap snipers were rendering us honors. Ducking down out of danger as swiftly as dignity permitted – or didn't permit – we turned our attention to a squad of Marines who were methodically trying to smoke some Japs out of a cave by lowering satchel charges from above the cave mouth. After the acrid explosion of one of these we could hear the occupants coughing. One young Marine sat with his rifle across his knees watching the cave entrance, rhythmically chewing tobacco. Occasionally he shifted his glance down to the road where our jeep was parked. We stopped to pass the time of day. He shifted his cud of tobacco, spat copiously, and nodded toward a small culvert on the road. "There's some Nips in that there culvert," he offered. "I've been watchin' them quite a while." Since our route back passed over this culvert, his remark was of more than passing interest. "Do you think you could hit one of them from here if he started to come out?" I asked. With a fine disdain for the two colonels before him, he spat again. "Yes," he assured us.

We regained our jeep to find that the driver had left his rifle in camp, not expecting to come so close to the front lines on this trip. I drew and loaded my .45, just in case. When Silver tried to follow suit he found that the slide of his automatic pistol was jammed with rust. I made pertinent comment regarding the old soldier's adage, "as dirty as an officer's pistol," as we spurred our jeep swiftly across the danger spot, protected only by one .45 caliber pistol among us, but confident in our Marine rifleman – toward whom we directed a wave of acknowledgement as we sped back down the West Road toward a quieter area, more appropriate to "headquarters types."

I might mention that while we were on the ridge we encountered the bodies of several freshly killed Japanese Marines who had apparently been barged down the Babelthaup the night before. These appeared to be larger, better-nourished men than those we had been accustomed to seeing. All their weapons and accoutrements had long since been lifted by the souvenir-conscious American Marines.

On September 28 we witnessed the well-planned and executed attack against the islet of Negesebus, lying adjacent to the north end of Peleliu. For this show the *Mount McKinley* had moved in very close to the fringing reef thus affording us a grandstand seat for the spectacle. The troop movement of battalion strength was scheduled to move in amptracs across the shallow, coral-studded, 200-yard inlet that separated the islands, after appropriate Naval gunfire and air support had softened up the defenses. We had planned a classic exhibition of beach strafing by "Cowboy" Stout's Corsairs to cover the last-minute approach of the amptracs. This operation, to be most effective, required split-second timing between air and ground units. In the event, the Corsairs arrived exactly on schedule with all guns firing parallel to the water line. I breathed easier as I shifted my glasses to the expected movement of the amptracs – which failed to materialize for five agonizing minutes after the last Corsair had completed his run. Fortunately the few defenders who were left alive chose not to come out and dispute the belated passage of the amptracs. Later we discovered the reason for the delay – a misunderstanding as to the signal for jumpoff. I regretted that such a perfect exhibition of close air support had to be even slightly marred – but then I was perhaps too much of a perfectionist in such matters.

About this time the weather changed as our operating area began to feel the effects of an approaching typhoon. Torrential rain and high winds began to seriously interfere with the unloading of supplies. Some of the lesser craft were swept aground on the reefs; our unloading pier of pontoon barges was repeatedly broken up. Finally the unloading had to be confined to LSTs operating from the somewhat-protected SE beaches, without the aid of piers. While the rains brought a measure of relief to the heat and thirst-plagued Marines ashore, the resultant logistic problems brought only headaches to the high command.

Early in October I accepted an invitation to return to Pearl Harbor with General Julian Smith and his staff; who, while nominally the senior command echelon for the expeditionary troops, had actually little to do insofar as Peleliu was concerned. We took off for Guam in an old DC-3 transport, on a course that took us uncomfortable close to the Japanese stronghold of Babelthaup. Although the Japanese air operations during the Peleliu campaign had been limited to nocturnal sorties of floatplane reconnaissance and minor heckling missions, we had no assurance that they did not have Zero fighters on Babelthaup. The apparent failure of our air commander to provide suitable escort for a senior general officer and staff was a source of momentary uneasiness – to me at least. However, I did not disturb General Smith with this thought. As the hours passed I noted some of the officers surreptitiously checking their watches. I knew of course that we were overdue on our arrival time – with no Guam looming up on the horizon. The weather was also closing in ominously. After an hour or so of increasing tension we made our landfall and touched down in the teeth of a violent squall that kept us sitting on the runway in a torrential deluge for fifteen minutes before we could disembark.

We spent a day on Guam, visiting old friends, touring the recent battlefields, marveling at the magnitude of new construction that had already changed beyond recognition the old Guam that I had last seen in 1928. We were then ferried up to Saipan for an overnight stay with Colonel Paddy McKittrick, one of my old Aviation cronies, who then headed the Saipan Air Defense Command. Next morning we transferred to a Navy DC-4, a four-engine transport, for the trans-Pacific flight to Pearl Harbor. The plane commander had only a neophyte copilot aboard so welcomed me to the cockpit for a turn at the wheel. We shortly ran into the fringe of the typhoon and for two or three hours suffered the roughest ride of my experience. With both of us fighting the controls to dampen out tremendous gusts that threatened to rip off our wings, we finally wore through the storm. I surrendered my seat to the young copilot and went aft into the cabin to check out our passengers. They were an unhappy lot, but after one look at my limp and sweat-stained shirt they relieved the strain by jovially commenting on the atrocious quality of my airmanship. We made our second refueling stop at Johnston Island in time for supper, then on to Pearl Harbor through the night. Thus ended the first adventure of the war.

In my report of the operation I was studiously critical of the Navy's conduct of air support, and gave due praise to what one squadron of Marine Corsairs at Peleliu had been able to accomplish. An Air Force general later quoted me at length in support of his findings that the Air Force also lacked the ability to provide true close air support for Army troops. His action did not find favor with his superiors in Washington. As for my original report, I suspect that it was suppressed at Pearl Harbor in the interests of inter-Service amity; I was never able to find it in the archives after the war. In any event I was able to apply the lessons learned to subsequent large-scale operations at Iwo Jima and Okinawa, so was always grateful for the opportunity to participate in the Peleliu campaign as an observer.

Shortly after my return to Ewa after this six-week interlude in the War Zone my new command was formally organized. The next four months were consumed in a continuing struggle to secure the specially trained personnel and the complex communications equipment which we would need to put the first landing force air support control unit in the field. While the basic radio vans and generators were available in local depots, all the necessary gear had to be designed, constructed, then field tested. Somehow, with the assistance of my brainy staff and group of technicians, this was accomplished – not without some abrasion here and there. My first priority task had to be persuasion of the brilliant but highly irascible Major General Pat Mulcahy, my immediate superior, that the future of Marine Aviation might well depend on how we were able to control our own supporting aircraft in the next big show. In the end I won out; with the weight of Pat's two stars behind me, depot commanders and personnel directors had no choice but to give me what I asked for, even to a special tank landing ship to carry our first unit into battle.

In December I was able to combine an official trip to the West Coast with a few days Christmas leave to spend with my little family in Coronado, whom I had not seen in almost a year. As I recall, the ostensible purpose of this trip was a conference with my old friend, Colonel Al Cooley, then at Santa Barbara training the first of his

124

Marine carrier groups. The time passed all too quickly before I had to catch a NATS plane out of San Francisco for Pearl Harbor; nevertheless, this was an unexpected and cherished break in my wartime routine, since few Marines were able to return Stateside until their combat tours were finished.

CHAPTER VIII

IWO JIMA – THE HOT ROCK

Some time after my return from Peleliu I had been made privy to the planning for the next operation, which would advance the Marines to the very doorstep of Japan's island empire. The assault and capture of Iwo Jima, a volcanic island in the Bonins chain of which few of us had ever heard, was intended to establish a U.S. air base from which escorting fighters could assist the B-29 bombers of the Strategic Air Force in their continuing efforts to knock Japan out of the war. These bombers, operating out of Saipan and Tinian bases, had been suffering severe losses due mainly to crippled planes not being able to get back to home base across 1500 miles of ocean. Then, too, the Japs on Iwo Jima not only gave warning of impending raids on their homeland, but also sent out harassing fighters to pick off stragglers. The situation for the U.S. Army Air Force (as it was then known) had become extremely critical; and so the Marines were dealt a hand in the game.

The Third Amphibious Corps, under Major General Harry Schmidt, with whom I had last served in Nicaragua some fifteen years earlier, was designated to receive the cards. I was told to have my first unit, LFASCU One, ready to load out with those elements of the Third PHIBCORPS which would stage out of Pearl Harbor, not later than the last week in January. D-Day for the operation was finally set for February 19, 1945. Meanwhile, I had completed the organization of my group headquarters, and had begun training the two additional units which were to be committed to the larger assault on Okinawa. I now elected to leave this second task in the hands of my executive officer, Colonel Ken Weir, while I took personal command of

LAPSCU One en route to Iwo Jima. While admittedly it was somewhat unconventional for a group commander to personally lead a squadron into combat, I felt I had sound reasons for doing so. I was the sponsor of an untried innovation in air support control tactics, responsible for its success or failure in this the first trial by combat. I could not reconcile myself to delegating this command to a subordinate, however competent.

The Iwo Jima attack force loaded and sailed from Pearl Harbor in late January, as scheduled; LAFSCU One, combat loaded on an LST, was part of the convoy. I still had work to do in Ewa, so did not plan to join General Schmidt's staff aboard the AGC *Auburn*, prior to the final rehearsal off Saipan. My flight log for February 1945 indicates that I left Ewa on the 3rd, aboard a Marine R5C (twin-engine Curtiss transport), flew to Kwajalein via Johnston Island, thence up to Roi on the north side of the atoll, where I found most of the ships of the attack force. I checked in with General Schmidt and staff aboard the *Auburn*, returned to Kwajalein, and left that same afternoon for Eniwetok. I spent the night with Colonel Dan Torrey, a one-time squadron mate in VF-9M, and the pilots of Marine Air Group 22, then operating off the island of Engebi against the Japanese-occupied islands to the southwest. I could only agree with them after witnessing a sortie against Ponape that this so called "milk run" against bypassed Japanese garrisons which had no capacity for harming us was but a futile and unnecessarily expensive employment of Marine Corps Aviation.

On the 7th we completed our flight to Saipan. My old friend and one-time flight instructor, Tom Cushman, had donned the stars of a brigadier and assumed the responsibilities of Air Defense Commander, Saipan. I spent several days as his guest, while looking over the Marine and Air Force installations on the island, which but a few months earlier had been the scene of one of our toughest battles. I recall that I spent one day with my friends of the Second Marine Division, then licking their wounds in a semi-permanent tent camp while reorganizing and retraining for their next big show – which turned out to be Okinawa. As I now recall, Colonel Jim Risely showed me over the battlefields that were still littered with the debris of war, and gave me a sound eyewitness account of what his Division had accomplished. I paid my respects to the peppery Division Commander, Major General Tommy Watson, with whom I had served in Tientsin in 1927, and then turned my attention to the Marine and Army aviators and their problems of air defense. I also found on Saipan a special squadron of twin-engine B-25 bombers commanded by Colonel Jack Cram, who in 1940 had been one of my young lieutenants in Marine Fighter Squadron Two. This squadron was equipped with special rockets for the night attack of Jap shipping, and was anticipating their later employment in the Iwo Jima area.

Since I knew personally so many of these Marine officers, both ground and air, most of ground and air, most of whom I had not seen since before the war started, these visits partook of reunions which helped bring me up to date on what had been happening in the Pacific these three years past. I did not fail to solicit comment on air support matters – some of which was pungently candid. The ground officers had not been happy with what they had; the aviators had been frustrated with the restrictions that had hampered their efforts to support their brethren of the gravel-filled footwear. It was all too apparent that both sides were demanding a change; as the catalyst I seemed to be very much on the spot. I could only hope Iwo Jima would be different.

Meanwhile, as we waited for the arrival of the attack force, I made arrangements with the depot people at Tanapag for the reception of LFASCU Two and LFASCU Three when and if they should arrive, and for the rehabilitation of LFASCU One after its return from Iwo. This last would be a rush job since this first of my air control units was also committed for Okinawa. In the actual event, however, the transfer at Saipan was accomplished easily since we got out of Iwo in ample time and with only minor material losses.

During the week that I stayed at Tom Cushman's headquarters camp, which was perched on a hill overlooking Aslito air base as well as the neighboring Ushi base on the northern tip of Tinian, I had the opportunity to witness the launching of a major B-29 attack against the mainland of Japan. The Marianas Wing of the 20th Air Force, equipped with these superbly designed and equipped heavy bombers, occupied the two bases mentioned above, each of which had parallel twin runways oriented generally east and west. The eastern end of these runways terminated at the edge of steep declivities that dropped off perhaps two or three hundred feet into the sea. Since the prevailing winds were from the east this layout of the runways obviated the necessity for the heavily laden bombers to climb for altitude immediately following takeoff, which was a fortunate circumstance considering their loads.

The bombers started taking off early in the afternoon, alternating on the twin runways at very short intervals. Although these runways were approximately 10,000 feet in length, each takeoff appeared to be a cliffhanger for the crew – as it certainly was for the observer! With the aid of my binoculars I could see that lift-off was delayed until the last possible second; in many cases the big planes deliberately lost altitude as they cleared the cliff, dropped out of sight momentarily, then appeared to be skimming the waters of Magicienne Bay as they turned gently northward. Since we knew that a loss of power on any one of the four engines would result in a fatal crash, the margin of available power over the weight of the aircraft being practically nil, we did not breathe easily until the last plane was safely aloft. A second group was taking off from Tinian at the same time, but due to the distance could not be so closely observed. The big bombers made no attempt to form up, but flew individually in a wide-open column as they slowly gained altitude and disappeared from sight. Presumably they pulled into squadron formation for mutual protection just before going into their bombing runs. Late that night the returning aircraft began to struggle in singly, homing on vertical searchlight beams as beacons. As I recall they all returned from this raid; it was not always – or usually – thus. Later, on Iwo Jima, I had daily confirmation of the hazards that these bomber crews ran on each and every mission.

Also based on Saipan was the bomber wing of the Seventh Air Force, whose commander, Larry Carr, had been a classmate of mine at Maxwell Field in 1936-1937. This unit was equipped with a B-24 medium bomber, which they had been using for the preliminary bombing of Iwo Jima and Chichi Jima – without notable success, I'm afraid. Larry gave me a tour of his operations department, which afforded me a better comprehension of our objective and what we would have to overcome to attain it.

I went aboard the *Auburn* as soon as the attack force arrived off

Garapan and reported to General Schmidt as air support commander, landing force. I found that I had a secondary assignment as deputy commander, air, Iwo Jima, to which I gave little thought at the time. The planned rehearsal appeared to be rather ragged – but then they always were. The air control circuits in the AGC control room functioned well enough, and the participating carrier air units answered signals satisfactorily. In any event we now had to be content with what we had; we were committed to what – had we then known – was to be an epic amphibious operation. On February 15 and 16 the armada weighed anchor and headed north.

I spent most of the next three days in the control room keeping abreast of the reported results of preliminary Naval gunfire and air strikes against our volcanic objective. These were, as usual, overly optimistic as to damage inflicted on enemy positions, and were taken with more than a grain of salt by General Schmidt's experienced operations officers. During this approach period I was also able to familiarize myself with the detailed plans for landing, and the topography of the beaches and adjacent areas. Even to my inexperienced eye we appeared to have drawn a tough nut to crack. This view was confirmed by the report on the 17th that Japanese coast defense guns had seriously damaged a number of our smaller fire support ships, including the cruiser *Pensacola*. Then on the evening of the 18th we learned that Kamikaze attacks had sunk one of our small escort carriers (*Bismarck Sea*) and seriously crippled the huge pre-war carrier, *Saratoga*.

The *Saratoga*, aboard which I had cut my teeth as a carrier pilot in 1939, was commanded by Captain "Fish" Moebus, an old shipmate af that earlier and more tranquil era. Thus I had a very personal interest in what was happening aboard my old ship. After a preliminary message that the "Sara" was under air attack, there followed an agonizing silence for perhaps a half hour or more. Then we were electrified by a flash message to this effect: "Heavy damage from Kamikaze attack x on fire from stem to stern and listing." Then came the next message: "Fires being controlled x will continue to land planes." We felt like cheering! "Fish" Moebus landed his planes in the gathering darkness and withdrew with honors from the combat area. His ship had to return to Pearl Harbor for extensive repairs but his pilots lived to fight again from other decks. (This is how I remembered it. The official reports say that the Kamikaze attack occurred on the evening of February 21.)

As I recall our last evening as we eased into the objective area off the eastern beaches of Iwo Jima, we had little reason to be cheerful. Obviously, the Naval guns had not neutralized the powerful Jap defenses; and we had every reason to expect devastating Kamikaze attacks during our landings on the morrow. Sometime during the night the vibration of the ship's engines ceased, and the gentle heaving of the vessel became more pronounced. For better or worse we had arrived.

We were on deck at first light. Around us were the vague silhouettes of more than 450 ships of the United States Navy, forming a crescent around the tiny island, which looked in the dim light like a tadpole with a large wart on the tip of its tail. Already, the troop transports were lowering boats, and amphibious tractors were sliding out the gaping bow doors of LSTs and forming their tight waiting circles. The cargo nets were draped over the sides of the troop ships as the Marines waited that last order to disembark.

Then at 0640 the supporting battleships, cruisers, and destroyers were suddenly obscured by clouds of dirty brown smoke, through which vivid flashes marked the muzzles of the main battery guns. A few seconds later we heard the rolling crescendo of thunder, followed by another as the crashing salvos obliterated the mist-shrouded outline of the volcano Suribachi, and the chaotic cliffs overlooking our landing beaches. This continued in a drum roll of sound that could be felt as well as heard for perhaps a half hour or more. Then the first waves of dive-bombers from the fast carriers came in over the beaches, leaving in their wake erupting volcanoes of fire and greasy smoke. These aircraft seemed to focus their attention on the lower slopes of Suribachi, and I recall that they did a precision job, pulling out of their dives at perilously low altitudes amid bursting anti-aircraft shells from the Jap defenders. I later verified that Marine squadrons were involved in this preliminary bombing; their daring apparently had rubbed off on their Naval colleagues, for all these pilots were superb.

At about this time, the *Auburn*, which could have been no more than a couple of miles offshore, took a light shell explosion high in her rigging – probably a stray from one of the Jap AA batteries. No

131

particular damage was done, save to the nerves of those of us who were on deck, but the whine of fragments over our heads was a reminder that a shooting war was no respecter of persons.

The Naval bombardment continued until 0805. Meanwhile the debarkation was well under way; the surface of the sea was marked by long white streaks as the landing boats began to move toward the beaches. Then the planes returned to pay their respects to the Jap beach defenses in a 20-minute attack which gave the fire support ships time to shift positions for the final pre-H Hour bombardment.

The finale of the air show for the Marines approaching the beach was a most spectacular razzle-dazzle performed by two squadrons of Marine Corsairs from the fast carriers, led by Lt. Col. Bill Millington, a one-time junior pilot in my pre-war squadron, VMF -2. Bill and I had discussed this beach strafing attack in Ewa during the planning stage. Regardless of what damage might accrue to the Jap defenders, we felt that the principal result of such a performance would be the lift in morale for the ground Marines approaching the beach – most of whom had never seen a Marine fighter squadron in action. I recall telling Millington that it would be necessary to "go in and scrape your bellies on the sand." That they did to the exuberant satisfaction of all concerned. No history of the Iwo Jima campaign fails to mention this beach-strafing maneuver. Happily, the Japs failed to hit a single plane of the group.

As Millington completed his last pass and pulled clear, the leading boat waves were but five minutes out. The bombardment ships immediately picked up the cudgel and began to walk their successive salvos inland, close over the heads of the landing Marines. For more than two hours now the landing beaches and adjacent areas had been subjected to such a pounding by Naval guns and air strikes that it seemed impossible that the defenders could still be alive. I must admit that I was tremendously impressed with this preparation, as were the other Marines involved. For once, the operation seemed to be going as planned.

However, the Jap moles were still there. I saw two of our gunfire spotter planes shot down right in the midst of the approaching boat waves, from altitudes so low that no escape for the crews was possible. At 0902, just two minutes late, the first wave of armored amphibious tractors hit the beach and waddled ashore. They were followed swiftly by the first wave of landing boats, from which Marines could be seen emerging in apparent slow motion as they struggled through the clinging black volcanic sand that covered the beach terraces. There appeared to be little opposition; this was later confirmed by the initial radio reports. Then as Naval gunfire shifted further inland the Japs began to emerge from their deep redoubts on the flanks of the landing beaches. Ugly black blossoms of smoke began to erupt from the sands of the beaches and among the increasing congestion of boats at the water's edge. The first waves of Marines were spared this barrage of mortar shells and were not pinned down until they had cleared the first high terrace; following waves suffered many casualties and much confusion resulted at the water's edge. It became sadly apparent that despite the magnificent preliminary bombardment, the Fourth and Fifth Marine Divisions were in trouble.

Nevertheless, before noon of this first terrible day the Marines of Harry Liversedge's 28th Regiment had succeeded in crossing the narrow waist of the island and were turning south toward Suribachi. From the *Auburn* I could see the left end of Harry's line facing south, still anchored to the beach. His supporting howitzers could be seen taking up their supporting positions and firing into the Jap defenses at the base of Suribachi – a surprising number of which seemed to have survived the air and Naval bombardment. On the north beaches the Fourth Division troops were also suffering heavily from flanking fire that came from caves on the right flank. We could see little of this area from the *Auburn*, but learned that the Division had some of their Sherman tanks ashore and were advancing doggedly. Casualties were reported heavy all along the front, and it was evident that the advance was not going according to planned schedules.

Throughout the long day the Naval gunfire and the air strikes continued on spotted targets of opportunity and in response to specific troop requests. The entire island seemed to be erupting at times in billowing clouds of smoke, dust, and fire. As darkness fell, the beleaguered Marines buttoned up their lines as well as possible and prepared for expected counterattacks. As at Peleliu, the darkness was relieved by the eerie light of star shells fired from the supporting ships, and by the occasional signal flares and rockets which streaked upward from Suribachi as the Jap defenders apparently besought supporting fire from their brethren now cut off from them in the northern end of the island.

The next three days were much the same. The 28th Marines slowly closed their vise on the lower slopes of Suribachi. The Fourth and Fifth Divisions had pivoted northward and after severe fighting had uncovered Airfield No. 1. The fast carriers of Admiral Spruance's Fifth Fleet had gone off toward Japan, leaving us but token air support from the escort carrier division of Admiral Cal Durgin. In the ensuing days we surely missed those valiant Marine squadrons headed by Millington and Marshall; the lightly armed Naval planes from the "spitkit" carriers were poor substitutes indeed.

During this period I spent most of my time in the operations room of the AGC, monitoring the air support nets and conferring with the staffs of Admiral Harry Hill and General Schmidt. While the Division commanders had been able to set up their CPs ashore, there was still no room for the more elaborate corps installations, which would include my own unit, LFASCU One. We could only wait with ill-concealed impatience to go ashore and get on with our assigned tasks.

At about 1015 on D plus four days, the leading Marine patrol fought its way to the peak of Suribachi. At 1020 a lookout on the *Auburn* spotted an American flag flying there. Electrified, we rushed for the telescopes. I recall nothing in my career quite so inspiring as that sight. Rosenthal's later staged picture of the flag raising ceremony has become as historically famous as the painting of "Washington Crossing the Delaware." Secretary of the Navy James Forrestal, who witnessed the breaking of the original flag, predicted that this incident would insure Marines against oblivion for the next thousand years. Americans of 1945, great and small, were not then ashamed to admit patriotic feelings.

Finally, at 0955, February 24, General Schmidt opened his command post ashore, defiladed inadequately behind a low sand dune on the western side of the narrow waist of Iwo Jima. We had gone ashore in a landing craft earlier that morning, picking our way with some difficulty through the wreckage of ships of boats and vehicles that lined the water's edge. To the northward only a mile away the battle still raged for the capture of Airfield No. 2. A little beyond the front lines the Japanese artillery still commanded what now passed for a rear area; our selected campsite seemed hardly compatible with prevalent doctrine for the location of a corps command post. We immediately started digging in.

Meanwhile I had arranged to beach the LST that carried all my equipment. The first vehicle off the ramp was our previous bulldozer, which proceeded without difficulty to tow our heavy van trucks off the sandy beach. Within a half hour we had spotted and sandbagged the vans and operations tent and by the end of the day had tested all circuits and were monitoring the air support frequencies. A camp of sorts had been made, with individual foxholes adjacent to the pup tents. My orderly-jeep driver, Corporal Thomas Saunders, located for us a dogleg section of an old Jap trench that was sufficiently narrow and deep enough to afford good protection. We delayed putting up my personal tent, nor did we bother with messing facilities this first day. K rations would have to serve for the time being.

We received our first compliments from the Jap artillery shortly after midnight that first night ashore. Whether they were deliberately seeking out corps headquarters, or were more interested in the battalion of 155 howitzers that with the usual perversity of corps artillerymen had sited their guns 100 yards in front of General Schmidt's tent, we neither knew nor cared at the moment. For exactly thirty minutes we were treated to repetitive salvos of high-velocity 4.7 shells that gave no warning of their approach. Although one of these exploded against the parados of our trench, showering us with dirt from the fragments entering the trench wall just above our prone bodies, Saunders and I escaped with but momentary concussion and temporary deafness. This was his very first experience under fire, so I felt impelled to caution him to keep his helmet on and his head snug against the trench wall. In the morning we found that one Marine who had been sleeping in a shallow depression had been killed instantly, and a few others had been wounded. Before night came again we became a city of moles; very deep foxholes were suddenly the compelling style, while our installations were double-sandbagged as high as the bags would stand. It was well we did this, for this periodic bombardment continued for several nights, always starting on a half hour and lasting for exactly thirty minutes. The only variation was the starting time. I recall one night that I counted twenty-seven explosions before falling asleep again. In military parlance this was known as "harassing fire" – an apt term. I did not move out of my ready-made foxhole into a tent for several days; not until that battery of 4.7s had been silenced.

After awhile we became accustomed to the incessant noise from our own heavy gun batteries flinging their shells north over the Marine front lines, as well as the answering crashes of Jap counter-battery fire coming our way. Then there was the occasional pulsating whistle of an enemy "spigot" rocket going overhead on a random trajectory that usually ended in the water off the SW coast. One night we were treated to a typical "Fourth of July" spectacle when a Jap mortar shell landed in the Fifth Division ammunition dump with most disastrous results in the way of explosions and pyrotechnics. Someone got a whiff of picric acid and started a gas alarm that caused considerable confusion before it was rescinded. The fires and explosions kept the southern end of the island alight until daybreak, and no doubt gave much aid and comfort to our enemy – if not to our own supply and ordnance officers. There was no such thing as a quiet and peaceful night on Iwo, but when men became weary enough they will sleep through any disturbance. So it was with us at corps headquarters.

I now discovered that the designated air commander for Iwo Jima, Brig. Gen. "Mickey" Moore, Army Air Force, had not arrived. As the designated deputy I was faced with the immediate problem of renovating the recently captured airfield, establishing some means of air traffic control, and servicing whatever itinerant aircraft we might be called upon to receive even before the field was ready. To handle these chores I chose one of my officers, Lieut. Comdr. "Ned" Broyles, a Naval aviator who had been assigned to my staff, gave him a radio jeep and an operator, and the responsibilities – if not the amenities – of an air base commander. Ned was a tremendous man, physically as well as otherwise; he could and did work 'round the clock, so long as we could keep him fed – he always seemed to be refueling from a K ration packet. He rounded up the Seabee detachment and soon had the graders and rollers going full steam on the pockmarked runways – with a fine disregard for the not infrequent shelling to which the Japs subjected his crews. Long before the runways had been resurfaced Ned was "talking down" crippled carrier planes and servicing on a regular basis the "grasshopper" squadrons of the Marine Divisions. Aside from a daily inspection of his activities I left him to his own ingenious devices.

Meanwhile, my operations officers were keeping up with the air support activities and gradually assuming control of certain phases. I managed to visit from time to time the command posts of the three Marine Divisions (The Third was now in the lines), to consult with their operations officers and air officers as to their needs. They were all less than complimentary about the quality of the air support received after the first few days – for which I had no remedy at the moment.

By the end of the second day ashore LAFSCU One was ready to take over all control of supporting aviation, but for reasons of his own the Naval air support commander afloat, Captain Dick Whitehead, U.S.N., was reluctant to surrender the reins. Finally, after the dispute had been referred to the highest echelons of command, we were informed that command would be passed to us at 0900, March 1st. This was to be the historic moment for the Marines, long anticipated and fought for. The Navy had now told us, "put up or shut up," and had predicted failure of our improvised communications equipment to duplicate the elaborate and permanent installations aboard the command ships.

My chagrin can be imagined when a vital generator engine refused to start on the fateful morning, and we discovered that the spare could not be mated to the multiple cable connections to our vans. We had to ask for a thirty-minute delay while the mechanics fumed and sweated with the balky engine, which had never before failed us. I even suggested to my chief mechanic, Master Sergeant Webster, that we cut the cable and jury-rig the spare generator to the radio vans. Webster, a telephone system maintenance chief in civil life, was most reluctant to cut his cables, realizing better than I the troubles that would ensue. In the end, I yielded to his better judgment; he got the generator going and we came up once again on the circuits, ready to assume complete control. Until we left the island, more than two weeks later, our equipment never again faltered – and our point was proved. I have always had a special place in my memory for Sergeant

Webster.

The battered and jumbled landscape of northern Iwo Jima made it extremely difficult to orient the inexperienced Naval aviators with relation to their assigned targets and our own front lines. Most of our forward air controllers were also so inexperienced and unreliable that we did not dare trust them with the actual direction of an air strike close to our own lines – as most of them were. Consequently, we rigged a couple of radio jeeps with remotes telephonic connections and sent our own operations personnel into front line positions to better pinpoint the air strikes. I felt it necessary to personally conduct some of these missions; and in so doing enjoyed sundry adventures.

We would spot the jeep behind whatever cover might be available, then carry our field telephone with us to a forward observation post, unrolling wire behind us as we went. Thus connected we could talk directly with our headquarters operations tent as well as with the flight leaders concerned. Since we were in the front lines we of course knew where they were, and since we could see our targets we could coach the diving airmen into the desired pattern. I found that the presence of a senior Marine aviator in one of these advanced observation posts inspired a certain degree of confidence among the troops. I also discovered an understandable reluctance among the rank and file to have anything in their immediate vicinity that might draw enemy fire.

One young artillery lieutenant, whose rock chimney lookout I shared one day as we controlled a strike by means of his telephonic connection with his firing battery, chided me about the rusty eagles on my combat jacket, expressing concern that a Jap sniper might recognize me as a bonanza target. He was visibly relieved when the mission was completed and I slid down the back of the chimney, rejoined the faithful Saunders who with carbine at the ready had followed us – against my orders – to this exposed position, and returned hastily to the comparative safety of our defiladed jeep.

On the morning of March 4th we picked up a distress call from a homing B-29 (Dinah Might) reporting ten minutes of available fuel and asking for landing instructions. Since Ned Broyles had only about 2500 feet of runway completed we were indeed in a quandary, since it seemed unlikely that a B-29 could get down safely and we could ill afford to have our one available runway blocked with wreckage. On

138

the other hand, the B-29 pilot had no choice. We briefed him on the approach and invited him to come in; he dragged that huge plane in on the propellers, only to discover at the last minute the telephone cable which the Signal people had that very morning, unknown to Ned, elevated some twenty feet across the end of the runway. Despite his critical fuel situation the pilot elected to take a wave off. Ned's control jeep was spotted at the junction of the two runways. The bomber's wing passed directly over it, carrying away the radio antenna as Ned and his helper dived for safety. On the next pass the pilot cleared the wires, stalled his big plane to a short landing and swung off the runway into the lee of a sheltering cliff as the Japs woke up and sent down a barrage of mortar shells. It was a remarkable exhibition of airmanship. His trouble it seemed was a faulty transfer pump that had denied him the use of his reserve tanks. We gave him enough fuel to get him to Saipan, and rather inhospitably suggested that he be on his way; the Japs would have liked nothing better than to have dropped a mortar shell on this monster that had just bombed their homeland.

This was but the first of many B-29s which for one urgent reason or other dropped in on us during the days that followed. One I recall came in with two dead engines on the one side, riddled with machine gun bullets, and with wounded men aboard. He, too, made a successful landing. Another landed with no brakes; he ground looped safely into a sandpit at the end of the runway, where we had to leave him for the while.

Our Marine and Navy air transport squadrons based in the Marianas now began making airdrops of ammunition and emergency supplies. After a few days of this Ned Broyles and his Seabee engineers had enough runway to accommodate these aircraft, but at first we restricted actual landings to the hospital evacuation planes, trying to avoid any concentration of aircraft which would draw down Jap mortar and artillery fire. The field evacuation hospital was set up at the base of a cliff at the north end of the field. We would bring in the evacuation planes on the northeast runway, taxi them under the shelter of the cliff for immediate loading and take off on the south runway. After the first flight or so these aircraft came in complete with their complement of Navy nurses. These truly angels of mercy were a source of concern to me, but they proved a great morale builder for the

139

bearded and dirty Marines lined up on their stretchers. Once this system was in full operation the elapsed time between receipt of a wound and a clean bed in the Naval hospital became a matter of six to eight hours. Fortunately, the Japs proved unable to interfere with our evacuation, and we suffered no accidents.

Meanwhile, of course, the main battle raged in the north, and my people were busy controlling the air strikes, such as they were. Some days the escort carriers seemed to have forgotten us, and we hated to face the silent reproaches of our badgered Marines. We knew of course that our Marine pilots were busy with Fifth Fleet operations against Japan proper; we didn't expect this explanation to satisfy our front line who sorely needed their support.

Then, on March 6th, there arrived on Iwo Jima a most welcome addition to our congested population. The 15th Fighter Group of Army P-51s landed two squadrons of their sleek-looking Mustang planes on our runway, and reported to me for duty. We fixed them up with a campsite on the western lip of the airfield and began an immediate appraisal of their air support capabilities. The pilots had no experience in the close support of ground troops; their planes were not stressed for dive-bombing; and their radios had only limited capabilities on the Landing Force frequencies. However, these pilots were anxious to try their hand at this new game, and more than willing to improvise with their equipment. As a result we settled on glide bombing, delayed action fuses on their heavy bombs, and control by LAFSCU One of the one frequency they had available – which was the one our air base commander had to use to control air traffic. To check their ability to answer signals and hit what they aimed at we scheduled a few practice runs off the west coast of the island. Impressed with their flight discipline and individual airmanship, we put them to work in front of the lines on March 8.

I particularly recall one P-51 mission that I personally controlled against a stubborn blockhouse complex in the Third Marine Division's zone of action close to their left boundary. I brought the planes in from the sea in a single well-spaced column, crossing the west beach in shallow high-speed dives. They leveled off parallel to our front lines and below 100 feet of altitude. Due to the high speed at release the bomb trajectories were practically horizontal. All except one struck at the base of the little hill on which the blockhouse was located, penetrated deeply into what proved to be an underground bunker and exploded. Fire and smoke emerged from the mouths of the connecting tunnels, one or two of which were behind our own lines. By the time the delayed fuse was activated the plane which dropped the bomb was at least a mile away; the next in the column was just beginning his approach.

One plane, however, dropped a little too high. The bomb struck the top of the concrete blockhouse at a very flat angle, then ricocheted like a skipped stone to land on another small hill several hundred yards away – which had just been captured after a hard fight by a platoon of our Marines. The next ten seconds were perhaps the longest of my life as I waited for disaster to strike. Thirty seconds passed, a minute, two minutes, and yet no explosion. I breathed a prayer of thankfulness – the bomb was a dud, and I later found that it had struck no one on the hill.

The P-51s were particularly useful in cleaning out the lateral gorges that obstructed our advance toward the northern tip of the island. They would literally fly into the mouths of these canyons, skip their 100-lb. bombs into the honeycombed cliffs, and zoom clear before the walls came tumbling down on the erstwhile defenders. I pulled one of these attacks for the 28th Marines on the west flank, much to the delight of my old friend, Harry Liversedge. The bruised and weary remnants of our frontline units perked up with cheers for the "doggie" airmen, and profanely wanted to know why the Navy planes couldn't use the same tactics. Their commanders were so pleased with this type of support that General Schmidt, on March 10, released the carriers from further support missions. The Mustang (P-51s) continued their support in the increasingly restricted enemy area until Iwo Jima was declared secure. Their timely and unscheduled assistance had certainly been an extra dividend, and the official Marine reports were generous in appreciation. Fortunately, none of the P-51s were lost on these missions; they were simply too fast for the Jap AA gunners.

In justice to the Navy airmen, it must be admitted that they were trying to do a tough job with far too few aircraft, of already obsolete types incapable of carrying the heavier armament necessary for the fortified positions they were asked to destroy. The pilots were largely untrained in the recognition of terrain features, and with but a hazy idea of ground tactics. Despite all this there were examples of outstanding performance, duly noted in the operational journals of the period. The escort carrier pilots did not escape unscathed from these missions; a goodly number were shot down, a greater number limped back to their carrier decks or forced landed on Iwo with heavily damaged planes. The Japs could and did hit them as they were attacking ground targets.

According to the official log, the designated air commander for Iwo Jima (Brig, Gen. "Mickey" Moore) relieved me on March 7th and assumed responsibility for air base operations and air defense. He did not, however, take away from my control the P-51s that I needed, so I spent much of my remaining time playing with these delightful new toys. Time was running out on me, however; I was scheduled for the Okinawa campaign, due to open April 1st. Accordingly, immediately following the flag raising ceremony on March 16, when the island was officially declared "secure," I boarded a Marine transport plane for Saipan, leaving LAPSCU One to embark and follow. As the transport cleared the runway and banked into a southerly course I looked for the last time on that smoldering volcanic rock, which I was never to see or care to see again. There remained and still remain searing memories impossible to so easily dismiss. Iwo Jima was and is the high spot in my forty years of military service. There were other campaigns and greater responsibilities ahead but none afforded me the personal satisfaction and professional pride that accrued to me by reason of having participated in this great amphibious epic.

A Tribute To The U.S. Marine Corps, by Ron Stark.
February 23, 1945…atop Mt. Suribachi, Marines of 3rd Platoon, Easy Company, 28th Regiment, 5th Marine Division, hoist the first American flag over Iwo Jima. As Old glory breaks at the skyline, a tremendous roar of shouts and whistles will emerge from the Marines on the beaches and the sailors at sea. Later this flag would be lowered and a replacement raised. Although the photo of the second flag would be cherished on the home front, the first flag belongs to the heroes of Iwo Jima.

CHAPTER IX

OKINAWA – THE LAST BATTLE

After three weeks ashore on Iwo Jima, Saipan appeared as an oasis of civilization where one could luxuriate in hot showers, clean clothes, and mail from home. I again moved in with Tom Cushman for a couple of days while I put my house in order, and arranged for the resupply and trans-shipment of LFASCU-1, which was following me in from Iwo bound for Okinawa. Then I flew over to Guam in a dilapidated Army DC-3 with matching surly crew – whose airmanship accorded with their attitude and appearance. I was relieved to get back on the ground.

I reported to the various Naval and Marine commands, observed the amenities and validated my travel orders. Admiral Kelly Turner, aboard his flagship, *Eldorado*, was then in Leyte Gulf, assembling the huge amphibious force that was scheduled to invade Okinawa on the morning of April 1st, 1945. Since I had orders to report to General Buckner, Landing Force Commander, I arranged a flight with the Naval Air Transport Service, leaving Guam at midnight on March 20 to arrive at Samar at daybreak. An Army shuttle plane dropped me off at Panuan, on the north side of the strait opposite Tacloban, from whence I endured a creeping boat ride across to the U.S. Naval Base Facility located at the mouth of the strait on the north shore of Leyte. The Naval port authorities were not particularly cooperative in providing jeep transportation for a transient Marine colonel (or so I thought at the time)), but I managed to locate the Marine liaison officer, who obliged me with a ride for the 20 miles or so that I yet had to go down the beach to reach my ship. I finally found a boat that threaded its way through hundreds of ships in that anchorage and deposited me aboard the *Eldorado* in time for dinner. I had spent twelve frustrating hours in traveling fifty miles, after a sleepless night. My opinion of the "rear area commandos" manning that particular line of communication would not have been construed as complimentary.

Once aboard the overcrowded command ship I was lucky to find a berth in a small stateroom shared by three other officers, an Army brigadier, a Navy commodore, and an Air Force colonel. We were not only integrated but entwined when it came to dressing in that restricted cubbyhole. We also had to share bathroom facilities with two elderly major generals who were billeted in the adjoining stateroom. As I recall, we all accepted the situation in good spirits and not without humor – it was all a part of going to war.

Since the *Eldorado* and the main transport group were not due to sail until the 27th, I had the opportunity to visit the Leyte beaches across which General McArthur's troops had stormed but a few months before. These were now almost hidden under a sprawling maze of Army, Navy, and Air Force installations that had been erected to support the recapture of the Philippine Islands. I also had the opportunity to study the reams of operations orders and annexes which would govern our coming operations at Okinawa. I recall spending one day ashore at one of the Leyte Beach airdromes that was occupied by a Marine Air Group, among whose officers I found several old friends from more peaceful days.

Upon my formal reporting to Tenth Army Headquarters for duty as air support commander, landing force, I unwittingly touched off a jurisdictional dispute between Army and Navy staffs. Captain Dick Whitehead, who was Admiral Turner's air support commander, insisted that I should have reported to him instead of to the Army command. He was backed by Admiral Turner who referred the matter to Admiral Nimitz for decision. The Army staff was understandably confused, and the designated tactical Air Force commander, Major General Pat Mulcahy, USMC, elected not to press the matter. In the end I reported to Whitehead, and the Landing Force lost its chance to exercise some initial control of its supporting aircraft.

Eldorado and the main transport group sailed at dusk on the 27th, with appropriate combat escort, for the three-day run to Okinawa. The slower flotillas of landing craft had pulled out on the 24th, while other elements of the amphibious force were converging on the Okinawa beaches from sortie points in Ulithi and Saipan, or so we assumed from the plans – we could not see them, of course. However, the ships that were in our company comprised a most formidable armada as they plowed inexorably northward for the final showdown with the Rising Sun; they ringed us round to the horizon, literally. As darkness fell that first night the more distant of the blacked-out ships faded into the gloaming, while those adjacent became only indistinct blurs in the night, holding their respective positions precisely through the magic of their sweeping radar. Not a light showed nor siren or bell sounded. Radio transmitters were silent for the voyage.

One knew that within those darkened hulls were crowded thousands of soldiers and Marines who were fated to cross the Okinawa beaches on that coming Easter morning. One could imagine them standing in little silent knots about the decks, or lying wide eyed in their crowded bunks in stifling troop compartments thinking the thoughts that men think on the eve of battle – particularly very young men who without previous experience could only conjecture what those beaches might be like.

We expected the landing to be tough indeed, opposed by fanatical, disciplined troops who would be defending the very portal to their homeland. We expected our losses on the beaches to be heavy despite all the preparatory bombardment by Naval gunfire and aircraft – which was even now proceeding apace as the shadowy convoy of transport ships moved ever closer to the showdown.

Those three days of our approach voyage blend in my memory into one. I must have spent many hours on deck during those nights, watching the ghostly forms to right and left, ahead and astern, thinking of what the immediate future might hold, breathing the crisp ocean air into my lungs against the necessary hours of sleep in our stifling cabin. During the daylight hours I was in the command room, keeping abreast of the reports coming in from the preliminary operations. These were optimistic – they always were – but the old hands had reservations.

The ships slowed in the pre-dawn darkness. Ahead could be seen the muzzle flashes of Naval guns, the arcing trajectories of red-hot projectiles, the glow of fires on a distant blur of shoreline. Above the hum of our ship's circulating blowers could be faintly heard the roll and reverberation of man-made thunder as our supporting battleships, cruisers, destroyers, and rocket gunboats increased the tempo of their shore bombardment.

Easter morning dawned clear with but scattered patches of obscuring mist, to disclose to the defending Japanese the hostile armada of some 1300 ships lying to or anchored off their western beaches. Against this overwhelming force they could launch only token raids of suicidal Kamikaze pilots flying a motley collection of obsolete aircraft. At 0650 the defenders were hit by massive air attacks from the U.S. fast carrier force; at 0700 the American troops began their debarkation; at 0800 the assembled and loaded boat waves started their 4,000-yard run to the beaches. The Naval bombardment lifted to higher ground inland, obscuring the ridges with smoke, blinding the Jap observation posts. A last-minute beach strafing attack by the supporting carrier aircraft pulled away as the first assault waves of our troops made good their landings, opposed only by desultory mortar and machine gun fire. It was difficult to see from the flagship exactly what was happening on the beaches, but to our amazement there appeared to be but little resistance; our troops reported back that they were crossing the beaches "standing up." For reasons of their own the Japanese had elected not to defend their beaches.

Wave after wave of our landing craft hit the beaches, unloaded, and pulled away to their mother ships for subsequent loads. Our troops, Army and Marine, were carried forward by sheer momentum to the high ground of the ventral ridge of the island, where they paused to consolidate and assess the situation. Behind them the tremendous buildup and congestion on the beaches continued, unhampered by enemy action. It had not been this easy on Peleliu or Iwo Jima.

Meanwhile, in the air control center aboard the *Eldorado*, much was happening. One of our destroyers mistakenly shot down one of our cruiser spotting planes, thus incurring a terrible blast from our irascible admiral. Sporadic Kamikaze attacks against our outlying picket destroyers were being reported, despite the vigorous efforts of our combat air patrols. As the morning progressed our radars anxiously scanned the northern skies for the expected major Japanese air attack against the sitting ducks of our amphibious fleet. Our fighter directors were vectoring combat air patrols here and there to intercept the lone intruders who did show on the radar screens. Our Naval air support controllers were busy assigning missions to the carrier strike groups continually reporting on station. Actually, I had no part in this activity other than as an interested observer; my turn would come a few days later when air control would be passed on to my waiting units ashore.

My three LFASCUS (landing force air support control units) all got ashore the first day and established themselves at the designated Army and Corps command posts. The unit assigned to Tenth Army headquarters found itself practically in the front lines of the XXIV Army corps with no Army headquarters to be found (the Tenth Army headquarters was still aboard ship and so remained for several days). The LAFSCU commander, Colonel Avery Kier, reported back rather whimsically that he felt rather lonely up there with the front line troops. I suggested that he pick out a more appropriate site adjacent to the tactical air commander's command post and await developments.

I went ashore the second day with Pat Mulcahy and Bill Wallace, tactical air commander and air defense commander, respectively, to choose sites for their command posts ashore between Yontan and Kadena airdromes. Avery Kier moved his unit into an adjacent farm site, which also became my headquarters ashore.

Meanwhile the ground forces had cut through to the eastern beaches of the narrow island, the III Marine Amphibious Corps turning north (left) while the XXIV Army Corps wheeled south, leaving as a common rear area between them a secure corridor for air and supply installations. So far resistance had been light and casualties few. We knew this couldn't last, so bent every effort toward getting our air control units in full operation before the troops would really need air support to cover their advances. For the first few days I commuted to and from the flagship, but when we were ready to assume control of air support activities, I moved ashore.

During this period I took time to observe the rural countryside and the plight of the native people. These had deserted their farms and villages at our approach, and had later to be rounded up and placed temporarily in concentration camps. They were then permitted to harvest the remainder of their trampled ripe grain and to collect what

148

was left of their scattered livestock. The tiny farms with their handkerchief-sized fields and small woodlands of umbrella pines were neat and picturesque. The houses were mostly of stone rubble construction, with peaked thatched roofs and board floors. Each house was surrounded with an outer garden wall of stone, which assured privacy as well as protection from cyclonic winds – as we later learned. The little fields contained barley, sweet potatoes, soybeans, and on the lower levels were diked to form rice paddies.

These refugee people were pitiable in the extreme, mostly old men and women and small children. The Japanese had taken all the able-bodied men into service or labor battalions, the young women for other purposes. When these frightened and starving people found that we meant them no harm their relief and gratitude were touching. They had been told by the Japanese that we meant to kill them, or worse. Day after day they filed down from hiding places in the hills in family groups, old and incredibly bent, and ragged beyond imagination. The very ancient were carried on the backs of their middle-aged sons, the little ones peeped shyly and wide-eyed from grandmother's skirts. As soon as time permitted we resettled them in their villages and farms and did what we could to alleviate their distress. War is indeed hell – and not only for the combatants.

While the first week in April was for the ground forces the calm before the storm, the U.S. Fleet that had brought them to Okinawa was enduring the full fury of the Kamikaze – the "Divine Wind." Wave after wave of nondescript Japanese aircraft, mostly manned by fledgling pilots, left their home islands to hurl themselves against the invading American ships. They had to run the gauntlet of the American fighter screen maintained by the supporting carriers; then the survivors had to pass through the hail of anti-aircraft gunfire thrown up by the outer and inner circles of picket destroyers. Finally, they were met by the fury of the final defensive fires from the battleships and cruisers protecting the main transport anchorage. Incredibly, some few of them survived to ram their targets in the center of the bull's-eye; others in desperation went for the first ships they saw, which were the destroyers on the picket circles. Our losses among these gallant but thin-skinned little ships were very heavy, indeed, and our Naval commanders were hard pressed to maintain their screening forces. Then, too, the successive waves of Kamikaze pilots began to saturate our fighter defenses; the carrier pilots were becoming exhausted. Had the Japanese Fleet been able to sortie at this point the issue might well have been in doubt.

Fortunately, some relief was at hand. The Seabee engineers had renovated Yontan airdrome within two or three days of Herculean labor. Marine Air Group 31, Colonel John C. "Tobey" Munn commanding, catapulted from two support carriers, arrived at Yontan on April 7th and immediately placed four squadrons (24 planes each) of Corsairs at the disposal of the air defense commander. The Japanese had ample reason to fear these gull-winged fighters, which they chose to call "the whistling death." The first 12-plane combat patrol took the air at 1750, and the next morning Marine fliers had "splashed" their first three Kamikazes. Henceforth, the Marine fliers took over the burden of the close-in air defense of the objective area, freeing the carrier-based units for operations closer to Japan. Operating from a land base in the center of the bull's-eye, so to speak, the Marine squadrons were able to maintain a very high availability and sortie rate. Their timely arrival had turned the tide for our air defense forces, and in the ensuing days and weeks the squadron operations reports abounded with accounts of interceptions and splashes; a goodly number of Marine pilots qualified as aces before the Kamikazes ran their course. After April 9th, MAG 33, under Colonel Ward Dickey, operating from Kadena airdrome, still under construction, joined in the fray and shared in the honors. Most of the Marine intercept missions were controlled from the air defense control center ashore, along with my LAFSCU-3, under Avery Kier, acting as backup control in the event of communication failure, which happened rather frequently at first. Consequently, I was able to observe these fighter defense operations in detail, with I must admit a vicarious thrill.

During most of April the two Marine Air Groups operating from shore bases were far too busy with vital air defense missions to render more than token support to the ground forces, which by this time had run into the main Japanese defenses. I had to pacify the troops with such support as I could get from the escort carrier group. This became increasingly difficult to accomplish as the average troop commander is mainly interested in the enemy in front of his lines, with only an academic interest in the air commanders' overriding responsibility for air defense of the overall area.

I personally witnessed several of the Kamikaze attacks, some while on the *Eldorado*, others from my command post ashore. Since every anti-aircraft gun in the fleet anchorage, as well as the shore batteries, opened fire on these low-flying intruders, the greatest danger appeared to be from our own gunners. On one occasion, while I was on the flag bridge with Admiral Turner, one lone Kamikaze pilot slipped into the anchorage and apparently selected the *Eldorado* for his target. He was hit in the final stage of his dive by a shell from one our own five-inch guns, which knocked off a wheel from his fixed landing gear, causing him to veer in course slightly and splash barely astern of our ship. He was so close that we could clearly see the goggled head of the pilot. Fortunately for us the bomb he carried did not explode. During this cliff-hanging episode the admiral seemed to enjoying himself immensely, alternating admonition and cheers for the defending gun crews as if he were a spectator at a ball game. As for myself, I was too scared to cheer and too proud to duck below decks.

During this same twilight period another Kamikaze got through to the battleship *New Mexico*, which was at the moment Admiral Spruance's flagship. The Jap plane flew into the side of the ship, penetrating the base of the superstructure to end up in one of the fire rooms. His bomb exploded, killing a number of the crew. This was the second experience of this kind for Admiral Spruance, whose previous flagship, *Independence*, had been shot from under him only a couple of days before.

I recall that Admiral Turner sent his superior a signal of courtly apology, welcoming him to our anchorage and lamenting our inability to protect him. Considering that only three or four of the attackers out of 150 or so had managed to pierce our inner defenses, I doubt that Admiral Spruance felt any negligence was involved.

A day or so later, again at dusk, I saw a single Japanese fighter plane flying at high speed over the anchorage at some 10,000 feet of altitude. He was maneuvering expertly to avoid the mushrooming shell bursts that pursued him across the still well-lighted sky. Suddenly he executed a graceful half roll into a vertical dive that ended between the stacks of one of our cruisers with disastrous effect. This was no neophyte pilot herded out of Japan on a one-way mission; this was an expert fighter pilot going deliberately to his death. I could feel only professional admiration and a twinge of regret for the end of such a warrior.

After April 16th the Kamikaze attacks gradually diminished into sporadic raids, but the Jap air force continued to harass us with individual night bombers, which in the aggregate did little material damage except to our allotted sleep time, but did provide unlimited opportunities for practice with our Army and Marine anti-aircraft guns; and in effect turned every night into a Fourth of July celebration. These gunners seldom hit anything – I can recall seeing but one intruder coming down in flames – but the falling fragments from their exploding shells perforated our tents and led to the construction of elaborate "sleeping foxholes" beneath a canopy of rough planks and sandbags.

My command post did serve as an inadvertent target for a visiting Navy fighter pilot who at the moment of his landing approach to Yontan airdrome discovered that the plane in front of him was Japanese. Our quick-witted flyboy tripped his landing gear switch, charged his guns, and chased the intruder down our company street at some fifty feet of altitude. Bullets from his .50 caliber guns raked through the thatched roof of the farmhouse which served as my command post, and caused a wild scramble for fox holes on the part of all hands. The stricken Jap plane turned toward the airfield and crashed in flames on the lip of the plateau. Fortunately, except for nerves, we suffered no casualties other than a perforated radio mast.

On another occasion, in broad daylight, I saw a crippled Navy fighter plane being escorted into Yontan by a Marine Corsair. As they banked over the beach at low altitude, with markings clearly visible, machine gun crews aboard the flotilla of small landing craft opened fire. The contagion spread to the gun crews among our service troops on shore, through which I was at the moment passing. I finally stopped the firing after some vigorous language, but too late. The stricken Navy pilot crash-landed into a rice paddy, without serious injury, reflecting no doubt on the ancestry and hospitality of his Marine hosts. General Geiger, III Corps commander, who also witnessed this addle-brained performance, reacted as only he could, stripping the offending units of their .50-caliber machine guns. The chief offenders, however, had been the crews of the small Naval craft anchored off the beach, who consistently fired shoreward without regard to where their missiles might strike. Both Marine and Army troops suffered casualties before those trigger-happy sailors could be curbed. Which adventures only prove the point that in warfare all hazard is not necessarily generated by the enemy.

Meanwhile the Marines were sweeping northward at a rapid rate. LAFSCU-1, headed by Colonel Ken Weir, which had been assigned to support General Geiger's corps, was hard put to keep up with this rate of forward displacement. My daily contacts with this unit involved longer and longer jeep travel over roads not entirely clear of Jap snipers. We provided such air support as the mountainous terrain and weather permitted, but experienced some communication and target identification problems that required some change in control procedure. On at least one occasion the attacking aircraft mistakenly strafed our own troops, inflicting several casualties. General Geiger, an old Marine pilot, took these mishaps in stride; I was deeply chagrined, of course.

After turning south, the XXIV Army corps, under General Hodge, ran up against the main Jap defenses and was effectively halted. Calls for close air support were multiplied. LAFSCU-2, under Colonel Ken Kerby, was in close contact with General Hodge's headquarters and well-informed as to their requirements. He had sold the Army on the virtues of close air support, Marine version, and now it was up to him to produce. We threw every plane we could scrape up into Kerby's zone of action and he fed them smoothly and effectively into Hodge's efforts. On one red-letter day (April 19) Kerby and his controllers put some 375 combat aircraft over the XXIV corps front, handling without incident as many as seven individual strikes simultaneously. This performance proved beyond further challenge the thesis of close air support control by the ground commanders on the spot, rather than by some distant Naval commander.

About this time the rains hit us with tropical intensity. Okinawa became one vast quagmire. The Jap road net, lightly constructed for their small transport vehicles, crumbled beneath the treads of our heavy trucks and tractors. The supply of troops in the front line units

153

became critical, and we had to resort to airdrops of emergency supplies – a task not without hazard for the delivery pilots.

My personal jeep, new at Iwo Jima, was literally worn out by my arduous daily travels. Fortunately, my good friend, Colonel "Rosie" Rosecrans, head of the local Marine supply depot, could be persuaded to issue me a new one. My driver, Corporal Thomas Saunders, was now able to smile again. In wartime even trivial blessings are appreciated.

The events described above were typical of my life on Okinawa. There were others, some exciting and dangerous, others trivial and amusing. Our living conditions were primitive and sometimes uncomfortable, but bearable; in comparison with life on the front lines we lived in luxury. There was too much going on all the time for boredom, and we were too busy to miss our usual diversions. I think most of us were well aware of the historical significance of what we were doing, and we felt satisfaction in being a part of it.

Due to its physical size, rugged terrain, and inclement weather, Okinawa posed for us a very different problem and in some ways more difficult problem than we had in Iwo Jima, for instance. One aspect of the problem was the magnitude of the forces involved. A field army, such as the Tenth, requires a multitude of supporting elements to keep two corps in line. When are added the air and Naval installations incident to such a prolonged campaign, we must have had at least 200,000 American military people on the island. Since I was not directly involved in the ground action, and my supervisory duties relative to air support did not normally extend beyond corps headquarters, my overall view of the fighting was more objective and less intimate than it had been in previous campaigns. I served in effect as a staff liaison officer between the Naval command afloat and Army and Marine commands ashore. My corollary command function as an air control group commander was primarily administrative, and could be largely delegated to subordinates. My operational responsibilities were supervisory, and did not involve the actual personal direction of supporting air strikes, as had been the case on Iwo Jima. Consequently, I had greater freedom in my contacts with higher commanders than would have been normal for an officer of my rank.

The actual front lines, however, were never more than a few miles away from my camp; the sights, sounds, and smells of battle were always perceptible. One of the military cemeteries had been located only a quarter mile upwind from my headquarters, where the dead were collected daily, prepared each night, and buried each morning in precisely located niches under a common mantle of earth. We found these necessary activities a depressing and melancholy reminder of the costs of war.

I did have frequent opportunity to visit with the Marine aviators who were based on Yontan and Kadena airdromes and to discuss with them our control techniques. The group commanders, Munn and Dickey, were old friends, and most of the squadron leaders were well known to me. I envied them their opportunities for combat leadership in the air, from which I was debarred by age and rank. I was also a frequent visitor to the adjacent headquarters of Pat Mulcahy, Bill Wallace, Ford "Tex" Rogers, Boeker Batterton, et al, who were concerned with tactical air command and air defense functions. Sometimes we had friendly differences as to the equitable division of aircraft between close air support and air defense functions, but usually the operational assignment of Marine squadrons was satisfactory to me – although I was never able to meet the full demands of the front line commanders (particularly Army) for more and more Marine air support.

After almost eight weeks of Okinawa, while the main battle was still being heavily joined in the southern ridges, I was directed to return to the rear staging areas to ready other units for the coming attack on Japan proper. I turned over my duties to my next senior, Colonel Ken Weir, and thumbed my way back to Guam on a Marine ambulance plane, working my way as copilot. The date was May 24th; that very night the Japs executed a daring and spectacular raid on Yontan airdrome. But for me, safe ashore on Guam, the war was over, although I did not then, of course, know that.

While I was on Okinawa much had happened. Rear Admiral Mel Pride, well known to me as one of the earlier Navy birdmen, had relieved Captain Whitehead as commander of the Navy's air support program, which now encompassed the Marine shore-based units as well. I had been assigned as his chief of staff (the only time in history that a Marine officer ever served as chief of staff to an admiral afloat) with additional duty as commander of Marine air support control units. Plans were afoot to greatly increase the number of such units, preparatory to the planned invasion of Japan. Best news of all, from my personal viewpoint, my new assignment was to carry the rank of brigadier general.

In the event, this promotion did not materialize. The then-director of Marine Aviation had proposed that I be relieved by a personal friend of his, already a brigadier, who had previously been relieved of his command in the Pacific, but was anxious for a chance at reinstatement. General Geiger, as senior Marine commander with the Amphibious Forces of the Pacific Fleet, effectively squelched this move, saying he could see no reason why I should lose the command that I had built up from scratch. His views prevailed, both with the Navy command and with the Marine Corps Commandant. I held my place at the cost of considerable enmity of the two disgruntled Marine general officers concerned, one of whom did nothing to expedite the wartime promotion I had earned, and for which I had been selected already by a non-statutory Board in Washington.

The *Eldorado*, once more in Guam, thus became my headquarters. I moved aboard and joined Admiral Turner's mess – a welcome change from field rations. Since my pilot qualification was somewhat in arrears I flew down to the Fleet anchorage at Ulithi with Colonel Ben Redfield, a round trip of some five hours in a twin-engine R5C transport plane. The airbase at Ulithi was commanded at the time by Colonel Frank June – as I recall – with whom we lunched. After a short indoctrination period in my new assignment, Admiral Pride suggested that I return to Hawaii, then to San Diego, for the purpose of inspecting our units in training in those places, after which I was to have fifteen days leave to spend with my family, then living in Coronado. Admiral Turner, in a generous mood, upped the ante to 30 days leave. I departed on June 4th, and after sundry flying adventures was deposited at the Naval Air Station, San Diego, on June 18th.

We had a delightful family vacation, spent partly under the healing canopy of the Big Trees in Sequoia National Park. I resolutely put aside all thoughts of war, past and future, and tried to concentrate on living for the moment. The time passed all too quickly.

I still had work to do at the Naval Amphibious Base at Coronado, which involved a fast trip to Washington, D.C. En route back to the Coast I spent a night with my father, whom I hadn't seen since I entered the war zone. I don't believe I told him about my prospects relating to Japan.

On August 2 I started my long journey back to the Western Pacific. After a short break in Hawaii, I boarded a flying boat for Manila, where the *Eldorado* was now anchored. While en route we intercepted the message that an atomic bomb had been dropped on Japan. By the time we landed in Manila Bay we knew that at long last the war was over – and that there would be no star on my shoulder, as if that really mattered.

We swung round the hook in Manila Bay for some three weeks waiting for the tangled skein to unravel. Mel Pride and I visited the ruins of Corregidor, the debris of old Manila, and the beautiful Crater Lake country to the south. On September 5 I said goodbye to my old Navy friends and boarded a flying boat for Honolulu. Here it took me ten days to wind up my association with the Fleet Marine Force, Pacific, (which I was destined to command in later years). On September 17 I joined my little family in Coronado, with orders to report to Marine Corps Headquarters in Washington.

We had to take our daughter, LaVerne out of the Bishop's School in La Jolla, where she had already spent a year, since none of us wanted to endure further separation. We were fortunate to get her admitted to Holton Arms School in Washington (from which she was to graduate four years later). My wife, Nell, and I settled down for an indefinite stay in the Fairfax Hotel – houses and apartments were not then to be had. We didn't mind; we were together and the war was over.

CHAPTER X

BETWEEN THE WARS

Washington – 1945-1946

Upon reporting to Marine Corps Headquarters on October 20, 1945, I found my first assignment was to be chairman of the newly convened Joint Army and Navy Board for the Standardization of Air Support Procedures. As I recall, the membership included two Army officers, two Air Force officers, a Naval aviator, and myself. The Air Force (then still a part of the Army) was the sponsor, so provided office space and secretarial assistance within their section of the Pentagon.

There followed a very lively discussion with gradual resolution of dissonant views, resulting after some two months in a unanimous report, which included a tentative operational manual covering joint air support operations. This report was accepted by all the Services (except for the Air Force, which noted certain reservations), and was the basis for subsequent changes in organizational and operational procedures within the respective air control units – some of which are still valid (1971). We heard that the Air Force members were in trouble for having to some of the provisions in this report. Possibly so at the time, but one of these officers, then Colonel James Ferguson, later became a four-star general, so his participation in the proceedings of the Joint Board could not have been too detrimental to his later career. I found the experience most stimulating, in that I was able to crystallize my recent experience in the Pacific war zone; and perhaps was able to influence my colleagues toward the Marine Corps

158

philosophy.

The Air Force authority to whom I reported for this duty was the then Major General Hoyt Vandenberg, whom I had first known as a junior captain on the faculty of the Air Corps Tactical School, where I had been a student in 1936-1937. Vandenberg later became Chief of Staff of the Air Force, and a member of the Joint Chiefs of Staff, which august body I was later to serve. As on many other occasions in my career I found this personal acquaintance helpful in dealing with him, then and later – even though Vandenberg typified the "wild blue yonder" philosophy of air operations, a view not shared by the Marine Corps.

Early in January 1948, I was reassigned to the staff of the expiring Army and Navy Staff College, and then engaged in forming the Joint Army-Navy Amphibious Review Board. This organization was housed in the New War Department Building (since taken over by the State Department) located in a river area of Southwest Washington, long known as "Foggy Bottom." This Review Board, comprised of some thirty or forty members of the various Services, who had participated in major amphibious operations during the War, all of whom had decided and often conflicting views. We spent six months or more reviewing the voluminous operations reports which had come out of the European and Pacific theaters of war, a tedious and laborious task. Finally, we came up with a somewhat ponderous "Tentative Manual for Joint Amphibious Operations," which was a distilled compendium of our experiences and sometimes-divergent views. Whether this report was ever formally approved by the War and Navy Departments I never learned. Undoubtedly it did influence subsequent directives and operational manuals.

The work involved in this compilation was done by committees. I was assigned as a matter of course to the committee on air support; later I was a member of the final editorial review committee. I must confess that I often found this work boring after having participated in some of the more exciting episodes of the late War. Nevertheless there were compensations; I could walk from the Fairfax Hotel to "Foggy Bottom" and thus was spared the daily traffic struggle. Furthermore, I was completely detached from the headaches of excessively rapid demobilization and consequent morale problems then being faced by the operational units of all the services. Reflecting on this turmoil, I felt more content with my quiet eddy. I was learning a great deal from my colleagues who had served only in the European theater; perhaps they also were learning something from the pacific veterans.

In July the Army and Navy Staff College was phased into the new National War College, and moved to Washington Barracks (now Fort McNair) into the impressive buildings formerly occupied by the old Army War College. I was one of the officers selected for the initial faculty of this new institute of higher learning. The first commandant was to be Vice Admiral Harry Hill, with whom I had served in the Pacific.

Meanwhile we had subleased from a traveling friend a comfortable furnished apartment near the old Wardman Park Hotel. This provided us with better living conditions at less expense, even though it did involve more than a half hour's driving morning and afternoon. Unfortunately, our friend returned earlier than expected, and we had to move into a house out in the Chevy Chase area. My commuting time was about doubled, but we did enjoy the expanded living quarters.

The summer was spent finishing the Joint Board Report, then with faculty and students assembled there began the fall semester's academic program. The first few months were devoted largely to lectures given by an impressive array of distinguished experts from various fields, political, economic, academic, and military. Most of these lectures I was able to attend, since my personal commitment was scheduled for the spring semester. This attendance qualified me for a certificate in "International Affairs." This was for me a pleasant and somewhat leisurely interlude. I recall serving as escort officer for various visiting celebrities, one of whom was the famous British socialist, Harold Lasky. Mr. Lasky arrived by train in rumpled tweeds, fully equipped in the English fashion with brief case and umbrella. He was flattered at his reception, and wondered how I could possible have recognized him among all those passengers!

Family life in our comfortable house continued to be pleasant, with an adequate but not overwhelming social life. I do remember particularly that I took our fifteen-year-old daughter, LaVerne, to the Army and Navy football game in Philadelphia, traveling by special train. The usual weather prevailed, and I've forgotten who won, but it was an enjoyable occasion for father and daughter.

In December, the selection boards met at Headquarters, Marine Corps. I was eligible for selection to brigadier, but well down from the top of the list. The fact that I had been previously selected by a wartime panel – but not promoted – did not entitle me to special consideration; I would have to run the gantlet once more. The then-director of Marine Aviation, Major General Field Harris, had persuaded the Commandant to specify in the precept convening the Board that at least one of the six selectees must be an aviator. I had no illusions as to who he had in mind – it was the aviator next junior to me in rank who was generally conceded to have the inside track. Fortunately indeed for me, the president of the Board was General Geiger, lately home from the Pacific, but ailing. I also had some staunch friends among the members of the Board. In the event, after an anxious two weeks, I was the aviator selected for promotion. There was some rather ill-concealed disappointment in certain circles, but there was no lack of joy in our household.

I was sworn in on Christmas Eve, by courtesy of my good friend, Colonel Bill Scheyer, who stayed late in his office to do me the honor. Then I went home to have my new stars pinned on by my wife and daughter, in the old tradition. I discovered tinsel stars on my house slippers and robe and a huge star adorned my bedroom door. My girls have always been fun loving. We had a wonderful Christmas.

In all the Services promotion to flag or general officer rank bridges a very considerable gulf – it is far the most important step in a military career. Less than ten percent of the professional officers who attain the rank of colonel can hope to be further favored. Such promotion opens the way to high command responsibilities, and also confers substantial perquisites and privileges – more appreciated than is the difference in pay.

In accepting this promotion I had been jumped over some sixty colonels, most of whom were old friends and companions of more than twenty years service in the Marine Corps. I could well understand their disappointment at having been so passed over, and anticipated some show of resentment over my own good fortune. To their everlasting credit, I must record that only a very few ever showed any change in their personal feeling toward me. As for my own feelings, I realized with no little humility just how fortunate I had been – the lightning had struck.

My Marine colleague of the NWC staff, Colonel Dudley Brown, also felt the lightning strike. Admiral Hill very graciously congratulated us, saying that he would have felt aggrieved if either of us had failed this test, but then had to admit wryly that he had no vacancies in his organization for two new Marine brigadiers. We were forthright directed to report to Marine Corps Headquarters for reassignment, with the stipulation that we return to NWC to deliver any lectures we had scheduled for the spring semester.

Norfolk – 1947-1949

I drew an assignment as chief of staff to the newly formed Fleet Marine Force, Atlantic, to be located at the Naval Operating Base, Norfolk, VA. The designated commander was Lt. Gen. Keller Rockey, whom I had first known as a major in Nicaragua, but with whom I had not previously served. Since this was to be a combined air-ground command echelon, the Commandant had that one of the two general officers assigned must be an aviator. There was no reason to believe that Gen. Rockey would otherwise have selected me as his chief of staff; quite naturally he would have preferred someone he knew better. In any event he accepted me with good grace and we got on well together.

I drove down from Washington on New Year's Day through the aftermath of a snowstorm, spent the night with an old friend of the Pacific campaigns, Brig. Gen. Merwin Silverthorn, at his headquarters at the Marine barracks, Portsmouth, Va. Next day I found Gen Rockey, one colonel, and a sergeant major, occupying borrowed office space in Gen. Silverthorn's Troop Training Unit headquarters at the Naval Amphibious Base, Little Creek, Va. This was the total on-board complement of Headquarters, Fleet Marine Force, Atlantic, a high-sounding but rather empty title for what appeared to be but a paper organization. Obviously, I had a job cut out for me.

In the ensuing weeks we were able to find more suitable quarters on the Naval Operating Base, proper, acquire the working nucleus of a field headquarters, and begin some exploratory effort to exercise a measure of control over our major subordinate units. Those were the Second Marine Division at Camp Lejeune, and the Second Marine Air Wing, at Cherry Point, both located on the coast of North Carolina, some 200 miles south of Norfolk. Both these units were then in full preparation for major amphibious exercises scheduled shortly for the Caribbean Sea, so we could do little to change the command arrangements. Gen. Rockey thought we should go along as observers on this expedition, so we sailed on the *Pocono* as guests of the Amphibious Force Commander, joined up with Admiral Blandy's flagship and the Atlantic Fleet, then at anchor at Port-of-Spain, Trinidad, where we endured several days of social activity and some sightseeing. These amenities concluded, we then sailed for the Vieques-Culebra area to witness the maneuvers. All this was very pleasant and instructive, but I didn't feel that our new command echelon had done much to weld together the ground and air elements of the Fleet Marine Force.

Back again in Norfolk, I managed to get a small transport airplane, with crew, assigned to our headquarters so that we could make frequent and easy visits to Camp Lejeune and Cherry Point. Our staff was augmented slowly as the selected personnel became available; gradually we came to make our presence felt, not only with the Marine units but with the Naval command echelons as well. In due time we were fully accepted as a type of command of the Atlantic Fleet, and were treated accordingly by Admiral Blandy and his staff.. By that time I felt that I had my job well in hand, and under the amiable direction of the convivial Gen. Rockey, was able to function with some competence as a force chief of staff. Life was not too difficult or too arid, except in one respect.

Since no personal quarters, other than a suite in the BOQ, had been available to me, I had of necessity left my wife and daughter in Washington. Then, too, LaVerne was completing her second year at the fine Holton Arms School, and we had no wish to disturb her. Consequently, our family life was limited to two weekends a month, when duty and the weather permitted me to fly up to Washington. This was only half living, but we had to reconcile ourselves to the enforced separation – one of the penalties of Service life.

Admiral Blandy had established his Fleet headquarters in the disestablished wartime Naval Hospital, located outside the Naval Base proper. There were several new brick residences on the grounds, including the one formerly occupied by the hospital commander. This was offered to me; and I made haste to accept, moving my family to Norfolk after school was out. Norfolk, thereafter, became much more desirable as a duty station, although I could never enthuse over its climate.

I had meanwhile selected my first aide-de-camp, 1st Lt. Harold G. McRay, a young Marine aviator from Kansas City. Although we couldn't have known this at the time, "Mac" was to be a member of our official family, off and on, for the rest of my active duty. He served me as personal aide on three separate tours of duty. All of us considered "Mac" as a member of our personal family; he filled the place of the son we didn't have, and our relationship was accordingly close over those varied years.

We also had to select and train our house stewards, black Marines who had been enlisted and supposedly trained as cooks and stewards. Our initial selections were not too happy but my wife was able to effect some degree of improvement; our ménage was at least passable – most of the time. The burden of this training was also shared by the aide who thereby acquired considerable facility in dealing with black stewards. As I recall, though, both my good wife and the aide shared some frustrations.

During mid-August, 1947, Gen. Rockey and I flew out to Camp Perry to support our Marine competitors in the National Rifle and Pistol matches. They did well, as usual, and we were pleased to see many old shooting friends among them. This was a welcome break in our routine.

In November, I flew up to the Naval Air Station, Argentia, N.F., with my operations chief, Col. Al Shapley, to observe the cold weather exercises in which Col. "Bo" Ridgley's regiment was then engaged. We found "Bo" camped out on the tundra, in weather that seemed no colder than we had left in Norfolk. This reality dispelled any illusions of parka-clad figures struggling through the arctic cold behind sledge dogs! I left "Bo" a copy of "My Life in the Frozen North" to while away his long evenings under the aurora borealis. On but one evening during our short stay were we able to observe this weird spectacle of flashing colored lights shooting up from the northern horizon.

Newfoundland impressed me then and later as a beautiful but inclement land, inhabited by a race of pinched and anemic people. Probably our wartime airbases provided them with some measure of prosperity – it couldn't have been much. I regretted that time did not permit my indulgence in some salmon fishing or big game shooting.

LaVerne had gone back to Holton Arms School in September, which event led to a commuting program between Washington and Norfolk. Among the three of us we managed to maintain the family ties, at some considerable inconvenience and no little expense. Since there was no comparable school in the Norfolk area, we felt the personal sacrifice well justified.

In February 1948, I again went to the Caribbean to observe the annual amphibious exercises. We had scarcely departed when Norfolk was hit by a freak late-season snowstorm which dumped a foot-thick coverlet over the Naval Operating Base. My family was literally snowbound for several days.

At this time both the Second Marine Division and the Second Wing were operating with drastically reduced forces due to the post-war erosion of personnel. Most of the enlisted men were lately out of recruit camp, so the level of performance of both Marine and Naval units were somewhat below the acceptable norms. We tried to make due allowance for these adverse factors, but could only in our memories recall the magnificent amphibious forces of which we had been a part during the Pacific War. It was difficult to feel much enthusiasm for these staged sham battles involving such attenuated and half-trained organizations.

My flight log for May, 1948, shows that I flew out to Camp Campbell, Ky., to observe some Army-Air Force maneuvers which stressed to employment of glider-borne and parachute troops. This was a precisely executed operation that I found quite impressive – and so reported. My pilot on this flight was 1st Lt. James Gibbons, who in later years served as my personal pilot in Korea. During the course of these exercises I made a short flight to the Air Force Base at Smyrna, Tenn., in an Air Force attack bomber (A-26), piloted by a youngish Air Force major general, W.D. Old by name, whom I had met somewhere during the late War. At the close of the maneuvers I was the dinner guest of General and Mrs. Percy Clarkson, along with several other dignitaries, including the then-Secretary of the Army, Kenneth Royal–with whose rudeness and tactless remarks I was not impressed. There was also at this dinner a slim, handsome, Air Force colonel, Benjamin Davis, Jr., who at the time was the senior negro officer in the armed services, a polished gentleman who was later to rise to three-star rank.

Early that summer we were offered one of the fine old colonial houses that adorned "Admiral's Row, with an unobstructed view of the entrance to Hampton Roads. These houses had been built for the Jamestown Exposition in 1907, after the commodious and elegant style of that period. Although the one assigned to us, the "West Virginia House," had been converted into a duplex, it was still more than ample for our purpose. The location, within a short walk from my office, was also much more convenient.

Shortly thereafter we acquired a new chief steward, who like the aide was to become a longtime member of our official family. M/Sgt. John Cole had just transferred from the Navy, where he had served as a captain's steward at sea for a number of years. He was shiny black, impressively large and dignified, and thoroughly competent in all his duties. Thereafter our household ran smoothly and well, until his retirement in 1955. He was indeed a most valuable member of my personal staff, greatly appreciated and respected by all members of the family. We kept in touch for many years after we both had left the Service.

In September I flew out to Cleveland to observe the National Air Races, in which I had participated as one of the Marine "circus pilots" on 1935. I found the scheduled events thrilling, but hardly as personally exciting as I had remembered them.

Again, in October, I was off to Eglin Field, in Florida, to witness some Air Force demonstrations, returning via Pensacola. I had last seen Eglin Field in 1937 when it had only turf and sand runways, tricky even for the P-12 Army fighter I was piloting on that occasion. The modern Eglin was hardly recognizable.

These excursions, together with the frequent and routine visits to Camp Lejeune and Cherry Point, broke the office routine and served to keep me current with the changing military situation. Colonel George Rowan, a longtime friend from Nicaraguan days, had meanwhile reported as deputy chief of staff, thus freeing me from too rigid adherence to my swivel chair.

Once more, February and March of 1949 found me engaged in amphibious maneuvers in the Caribbean area. McRay and I flew the Beechcraft transport from Norfolk to San Juan, P.R., via Guantanamo Bay, Cuba, and Port au Prince, Haiti. There appeared to be little change in the Haitian capital city since my last stop there in 1935; in fact, not too much change in the 25 years since I had served in Haiti. On the return flight a month or so later, we refueled at Ramey Air Force Base on the western end of Puerto Rico, and were thus able to overfly Haiti on the way to Guantanamo. We took advantage of this flight to skirt the north coast of the big island for a nostalgic view of Cap Haitien and Plain du Nord, where I had spent two very interesting years of my military youth.

Some months before I had been told that I would be ordered to Guam the following summer to command the Marine Air Group 24 (Reinforced), at that time the only Marine Air unit still operating in the Western pacific. My reaction to this news was "singularly unenthusiastic," according to the then-Director of Marine Aviation, Major Gen. Bill Wallace, who chose to take some personal offense at my not-abnormal reaction to what I considered a most undesirable assignment. The matter was resolved by returning that lost air group to the United States.

In April 1949, my orders were changed to command the reinforced air group then operating from the Marine Air Station at Edenton, N.C. I didn't consider this assignment much of a professional plum, either – in fact I had begun to believe that there might be a punitive element involved. Nevertheless, I accepted the situation in good faith and made preparations to move down to Edenton and occupy the ante-bellum mansion that was to be our living quarters.

Washington – 1949-1951

Somewhat to my amazement I received a call from Wallace early in June advising me that the Commandant, General Cates, had that I

167

should come to Washington as Assistant Director of Marine Aviation. I was reasonably certain that this change had not been originated by Wallace.

Meanwhile, our daughter, LaVerne, had graduated from Holton Arms, *cum laude*, and had returned to Norfolk to summer with us. Now we would be going back to Washington early in August, while she would be leaving us in September for Wellesley College. Her cousin, Alma Ann Clinkenbeard, of Oklahoma City, had come to visit the preceding summer and had decided to attend Holton Arms for her last year of high school work. The two girl graduates made a lovely pair at commencement exercises, and we of course felt very proud indeed.

Gen. Rockey left early in June for his new assignment, to be relieved by Lt. Gen. Roy P. Hunt, an agreeable gentleman and competent professional soldier whom I had known for at least 25 years. I stayed on with him for a couple of months, to be relieved in turn by my old friend, Frank Schilt. By August 10, I had briefed Frank in his new duties, and we had moved into our new apartment at 2311 Connecticut Ave., in Washington. On August 18, I reported for my new assignment.

Although Washington duty deprived me of the amenities of living on a military base, I nevertheless was pleased to be returning to duty with Marine Corps Aviation for the first time since the War.

I had barely settled myself in my new Pentagon office when Bill Wallace took off for an extended tour of Europe and Africa, leaving me holding a rather limp sack. In one week in September alone, I was called on to make three public appearances, ranging from New York to Plymouth, N.C. – which diversion under the circumstances I did not find helpful.

Then in early October the Marine Corps was precipitated into that infamous inter-service controversy to be known as "The B-36 Hearings," in which a lot of dirty linen was being washed in public. Admiral Bedford, who was quarterback for the Navy team, asked that I be made available to his group to represent Marine Aviation. Consequently, I had to drop everything and prepare an appropriate presentation of the Marine Corps position, to be delivered in person before the Armed Forces Committee of the House, then headed by the venerable and irascible Carl Vinson. This document, of necessity, was very critical of the Air Force record during the late War, insofar as the effectual close support of Army ground forces was concerned. I expected to be grilled, and was prepared to fully document my statement before the Committee. Aside from some glowering and murmuring, however, the Air Force witnesses elected not to challenge me, and my statement went into the record verbatim. General Cates later took the stand and backed my statement as representing the Marine Corps position.

As a direct result of these hearings, Admiral Denfield, then CNO of the Navy, was summarily relieved of his assignment, and General Cates was threatened by the Secretary of Defense, Louis Johnson, and the Secretary of the Navy, Francis Matthews, political appointees both, with the same fate. However, General Cates had very strong backing in Congressional circles, so much so that the civilian secretaries did not dare carry out their intention. The retention of Cates made my own position safe, although I had fully expected to be forced to retire. I had reconciled myself to the position that if the Marine Corps was to be emasculated and reduced to ceremonial functions and Navy Yard guard duty – as the Army and Air Force seemed to be proposing – I did not wish to continue on active duty.

In the end, however, Congress took a firm stand and prescribed by law that the Marine Corps should not be reduced below a combat strength of three Marine divisions and three Marine air wings. Furthermore, the Marine Commandant was to be on an equal footing with the Navy CNO insofar as his administrative relations with the Secretary of the Navy were concerned; and was also to have a voice in any JCS proceedings that might affect the Marine Corps. Those of us who had put our careers on the line, so to speak, during the Congressional hearings were fully vindicated – and perhaps gained a measure of increased respect among our contemporaries. In my case, I had been permitted to participate by default; the honor and risk should have gone to the Director of Aviation – not to his deputy. However that came about, such prestige (or notoriety) as may have accrued to me during those proceedings certainly did no harm to my future career.

In November I flew out to Kansas City to address the American Legion on Armistice Day, one of several speech-making jaunts that came my way at the request of the sponsors concerned. Aside from these brief interludes, however, I was scarcely able to leave my desk all winter.

As an aftermath of the B-36 controversy, in an apparent attempt to heal the wounds, the Joint Chiefs of Staff decided to widen the inter-service membership on the Joint Staff, which hitherto had not included a Marine general officer. They asked General Cates to nominate a brigadier for assignment as the deputy director of the Joint Staff, for intelligence matters (G-2). My name was submitted, and on January 14 I reported to Rear Admiral Art Davis, then director of the Joint Staff, whom I had first met under rather painful circumstances when I was a student flyer at Pensacola and he was superintendent of training.

I had no previous experience in the intelligence field, whatsoever, but found that my department was well staffed with military and civilian experts. In due course I absorbed enough of the technique and vernacular to be able to represent my department in conferences and presentations. Perhaps my previous lack of experience proved to be an asset; in looking at the forest I was never bothered by the trees.

The most significant part of my duties was the daily briefing of General Bradley, then Chairman of the JCS; later I had the privilege of appearing before General Marshall for weekly briefings, after he relieved "Lord" Louis Johnson as Secretary of Defense. During the six months or so that I served under Johnson I was never in his office – such trivia as military intelligence did not appear to concern him. I found both General Marshall and General Bradley very considerate and tolerant gentlemen, professionals of the highest rank whom I could respect. I consider my service under them one of the highlights of my military career.

As a member of the Joint Intelligence Board I also sat in on weekly conferences with the Director, Central Intelligence Agency, then the peppery but most astute Gen. "Beetle" Smith, who had been Eisenhower's chief of staff in the European theatre. Other members of the Board were the chiefs of intelligence of the State Department, the FBI, and the individual military Services. Our conferences were never dull, often quite stimulating; and although our presiding officer, Gen. Smith, was ruthless with his own staff, he was always courteous and considerate with his colleagues on the Board. I learned a great deal more than I contributed in these sessions.

In June of 1950 I flew over to Frankfurt with my civilian deputy, Dr. "Pete" Craig, to attend the annual conference of the military and naval attaches accredited to our European embassies. Later we visited Trieste, Rome, Paris, and London before returning home. While in Rome I flew down to Palermo, on Sicily, for a courtesy call on the Sixth Fleet commander, Vice Admiral Ballantine, an old Naval aviator of long, if distant, acquaintance. From Rome I flew back to Frankfurt for a conference with the commander, Naval Forces, Europe, Rear Adm. John Wilkes, whom I had not previously met, but who was most hospitable during my stay in Heidelberg (which was the headquarters of the American European command).

On the flight from Rome I piloted the Naval attaché's Beechcraft across the main ranges of the Alps, an easy task compared with my experiences in the much higher Peruvian Andes. Incidentally, this same attaché, a Commander Thompson, had also served with me in Peru. I might also mention that I was the overnight guest while in Trieste of General and Mrs. Hoge, who were quartered in the old castle Miramare, once the home of Maxmillian and Carlotta. Although this ancient castle had been the headquarters of several conquering armies, the original furnishings were intact – a most unusual circumstance.

From Frankfurt I flew to Paris to rejoin Pete Craig for a little sightseeing, which included an evening at the Lido celebrating my fiftieth birthday. Then we flew over to London, made our manners to our British friends, and were invited to witness the King's birthday celebrations. This bit of majestic and impressive pageantry ended with a military exhibition involving horse-drawn artillery galloping around an arena with what appeared to be suicidal intent. No Roman chariot race could have been more thrilling. Fortunately, but unusually as we were told, there were no major casualties.

This was my first trip to Europe. I was appalled at the war damage still unrepaired in Germany, and impressed with the clean-up program well under way in London. Paris and Rome had been largely spared, we noted, but we had no opportunity to visit the battlefields of France and Italy. One highlight of my visit to Germany was a Sunday excursion down the Rhine from Mainz to Coblenz and return. Hitler's private yacht had been commandeered by the American Army as an excursion boat, so we traveled in luxurious style along the fabled river, complete with German band, my personal one-star flag at the masthead. As I recall, there was little war damage apparent except at Coblenz and a ruined bridge or so along the way. We had a long flight home from London via Shannon in a Boeing strato-cruiser, and then reluctantly returned to our desks.

Aside from the pleasures noted we gained much from this trip. We came to know personally most of the American commanders and their principal staff officers, as well as the intelligence personnel whose activities were of prime interest to us. One who stands first in my memory was the redoubtable "Iron Mike" O'Daniel, an Army major general of the old school, then assigned to our embassy in Moscow. Mike's philosophy, as expounded to the conference, was to the effect that the Russians were not really ten feet tall, and that their armed forces were not as invincible as some of us might have been led to believe. This was a refreshing antidote to some of the contemporary prophets of doom. In later years Mike headed up the Military Assistance Group in Saigon; after that I saw him once more, now retired in San Diego. He was indeed a colorful character, and I esteemed him as a friend.

Later in the summer my department hosted a visiting British delegation from their Joint Intelligence Group, headed by a delightful and convivial major general, by name Peter Shortt. We held a lengthy joint conference, exchanged views, on procedures for exchanging intelligence data, and generally promoted international amity. Peter was an American Civil War buff, so I arranged a car, chauffeur, and personal aide, for a three-day tour of the Virginia, Maryland, and Pennsylvania battlefields. He was delighted and grateful for this opportunity; we parted firm friends, maintaining a personal correspondence for some years later.

The rest of my time on the Joint Staff was more or less routine, sharpened by our entrance into the Korean civil war in the summer of 1950. The outbreak of hostilities, which was transmitted to me by a telephone call at about 4:00 A.M. on a Sunday morning from an alert and somewhat excited duty officer, caught the U.S. Intelligence Community almost completely unawares. We had, it seemed, depended almost entirely on the reports from General McArthur's headquarters in Japan, whose intelligence officer must have been wearing rosy spectacles most of the time. In any event the Joint Intelligence Staff had no laurels on which to rest.

During the first year of this current Washington tour I had undertaken some extra-curricular activity that cut rather heavily into our social and family life. I had enrolled in George Washington University for a course in Russian History, and as an afterthought, a second course in advanced Spanish. I managed to conclude these courses for credit, took some additional advanced standing examinations for further credit, and thus qualified at long last for the bachelor degree from Oklahoma A&M College, the institution I had left in 1919 to join the Marines. The arrival of my diploma was an occasion for family celebration.

Meanwhile our daughter, LaVerne, had her coming out party at the Army and Navy Country Club in Washington, and had completed her freshman year at Wellesley. Her welcome presence at home during the summer and the holidays certainly livened our household with youthful gaiety, even though it might have on occasion strained the single bathroom facilities of the family cliff dwelling.

Cherry Point – 1951

In July 1951, I again found favor with my selection board and early in August was promoted to major general. I used this advancement as a lever to spring myself from the Washington scene for a more appropriate command assignment. On August 13 I reported to Cherry Point, N.C. as relief commander of the Air Station and outlying facilities. The quarters normally assigned were still occupied by my retiring predecessor, an old friend whom I did not wish to further embarrass, so I took a suite in the BOQ, leaving my wife and daughter in Washington for the time being.

My friend and former flight instructor, Tom Cushman, also a recent major general selectee, had already assumed command of the Second Marine Air Wing. I would have preferred his job, but was glad enough to have a command of my own after six years. I had never commanded a Station so had to learn how to handle 4,000 civilian employees, their labor unions, and their congressman. It was an educational experience! I managed with the help of a competent staff to provide the services required by the operating groups, so that Tom Cushman and I had no problems in cooperation.

We were settled in our new quarters by November, in time for me to catch up on my hunting and fishing, and for us to plan the Christmas season, when once again the younger set would be dashing in and out of doors. We settled down for a tranquil year, with our reassembled domestic staff and smooth-running household. It was not to be.

El Toro – 1952

Early in January I was ordered on temporary duty to Sandia Air Force Base in Albuquerque for an indoctrination course in atomic weapons. Scarcely had I returned when I received dispatch orders to proceed to our El Toro Base in California as relief for the ailing Bill Wallace, who was facing retirement. The new assignment, as commanding general of Aircraft, Fleet Marine Force, Pacific, was more to my liking, a definite promotion, so we packed up again without too much regret and headed for the West Coast. I left Nell in Oklahoma, temporarily, while I hurried on to El Toro. Wallace had already relinquished command to his deputy, Colonel P.K. Smith, whom I formally relieved. I had agreed in Washington not to disturb Wallace in the occupancy of his quarters pending his full retirement. Since what I considered a more desirable house was available on the adjacent Santa Ana station, I did not begrudge Bill his final months of amenities.

Nell flew out within a few days, our household goods arrived, and we managed to get unpacked just in time to receive the new Commandant, General Lem Shepherd, and an accompanying congressman of the same name. Nell did herself very proud indeed with a formal dinner party arranged without notice for a household but two days established. The Commandant was very complimentary, then and later; the congressman – as was too often the case – took all the attention as his just due, and didn't bother to acknowledge the courtesy. Notwithstanding this sour note we were off to a pleasant year. McRay, now a captain, was back with us again. Sgt. Cole had rejoined us as at Cherry Point; and we had acquired a new cook, Sgt.

Charley Meyers, who proved to be a jewel in the culinary department, and who also stayed with us until we retired in 1959. We enjoyed our pleasant home, our privacy, and the social life of the community, civilian and military. I even found the six-mile daily drive to El Toro Base, where my headquarters was located, a very pleasant journey through the bean fields and eucalyptus avenues of the Irvine Ranch. We declined to move when the Wallace's left, yielding our claim to that set of quarters to my deputy, Frank Muir.

I had scarcely settled down to normal routine when I had to fly back to Washington for the quarterly session of the Navy Air Board, of which I was now an ex-officio member. Aside from the benefit of such meetings with the Naval Air Establishment, the cross-country flights involved opportunity for brief overnight visits to relatives living in Oklahoma.

The overriding mission of my new command was to provide personnel and materiel support for the First Marine Air Wing, then operating in the Korean war zone, where they had been for more than a year. I was anxious to get out there personally to appraise the situation, since I had never been to Korea. In early April I flew out with Major General Jerry Jerome, the designated relief for Frank Schilt as wing commander. Major General Bob Pepper also accompanied us, since he was later scheduled to take the Third Marine Division into Japan as a reserve force. This long over-water flight via Honolulu, Wake, and Japan required some forty hours of actual flight time each way in an unpressurized airplane – not exactly a pleasant trip.

I was pleased to renew acquaintance with the Honolulu area after an absence of six years, but was particularly interested in our brief nocturnal visit to Wake Island, where my old fighting squadron companions had exhibited such valor – and where so many of them had died.

This was also my first visit to Japan, where we passed a couple of busy days in Tokyo calling on the various U.S. commanders, including General Mark Clark who had relieved General McArthur as United Nations Commander in the Far East. We then flew over to Korea and spent five days visiting all the Marine air units, as well as the front-line troops of the First Marine Division, then commanded by Major John Selden, well known to me, of course. We were flitted around his front-line positions by helicopter, getting a first-hand – if sketchy – impression of the Korean War. Oddly enough, what I remember best from that trip was the beauty of the pink azalea shrubs, just coming into bloom, which carpeted the otherwise barren mountains. We returned to El Toro via Japan, Midway, and Honolulu, having crowded considerable and indoctrination into just fourteen days. I felt, however, much better qualified to realize the needs of our flying Marines on Korea's bleak air bases.

Meanwhile, at El Toro, we had uncovered a scandal involving a senior NCO charged with selling passes to gullible new Marines, and other irregularities. I had ordered his trial by general court martial, which dragged on interminably due to the delaying tactics of his civilian counsel, a radical lawyer from the legal fringes of the Loc Angeles area. Eventually we were able to have this lawyer convicted in Federal court for conspiracy to steal government property (one of his henchmen had actually burglarized the files of our legal office), but not before he had written scurrilous letters to everybody of consequence in Washington, accusing me and my legal staff of gross malfeasance of office, or worse.

Certain members of Congress were initially taken in by this smear campaign, resulting in some rather pointed inquiry from the office of the Navy Judge Advocate General, and the Marine Corps Commandant. Fortunately, our actions stood up under legal scrutiny; the congressional inquiries suddenly dried up when their authors were informed by some of their influential constituents, which included the local judiciary, that the originator of the smear campaign was not only an avowed Communist, but now a convicted felon. Nevertheless, I did not welcome all this notoriety in Washington circles, which might well have affected my career adversely – particularly had it come a bit later. This cause célèbre occupied a disproportionate amount of my time and effort for several months, and caused me no end of worry and annoyance. In the end the original court martial was reversed by the Court of Military Appeals, on some technical point or other; nevertheless, we had been able to clean up an unsavory mess – which I had inherited.

After I had returned home from the June meeting of the Navy Air Board, and had welcomed our daughter home for the summer, I asked

176

for a modest ten days leave. We drove up the coast route through Northern California, cut across to Portland and Seattle, and then crossed over the border for a brief look at Vancouver and Victoria, returning via the Olympic Peninsula and the Sacramento Valley. This was rather an ambitious project for a ten-day period, but we found it an enjoyable change of pace – with an opportunity to renew our close family relationships.

I was recalled to Washington late in July to serve on a selection board – colonels to brigadier. Major General Oliver P. Smith, of Korean fame, then commanding Fort Pendleton, accompanied me. Major General Graves Erskine, the "Big E" of some of the Pacific campaigns, was President of the Board. The precept of this Board was so worded by the Commandant as to permit more than the usual latitude of selection, and the rumor got around that we were expected to be well down the list for a particular officer of alleged brilliance, well known to be a protégé of the Commandant's. The Board, in its deliberations, could find no justification for passing over more senior colonels of good war record; and, perhaps feeling some resentment with this insidious administrative pressure, made its selections in accordance with conventional standards.

In any event the list presented to the Commandant failed to please him. He could have legally disapproved the findings of the Board and convened another, more amenable perhaps to his ideas; or he could have forwarded the report to the Secretary of the Navy with his disapproving endorsement. He did neither. He approved and forwarded the report, and then called in nine general officers that were members of the Board for an unprecedented and most unjudicial tongue lashing, which left us speechless with shocked incredulity. The idol of the uncritical had shown his feet of clay.

No member of that Board could expect any special consideration afterwards from Headquarters. In my own case there was thereafter a feeling of uneasy restraint, possibly unjustified, in my relations with Gen. Shepherd. That feeling persists after thirty years.

In late October I accompanied Lt. Gen. Frank Hart, who, as Commander Fleet Marine Force, Pacific, was my immediate superior, on a second trip to Korea, ostensibly to check on the current needs of Jerry Jerome – whom I hoped to relieve at the end of the year. Again I visited all our airbases and the front line Marines, noting the autumnal chill that had already begun to seep down from the northern mountains. On this trip I took the opportunity to indulge in some fantastic duck and pheasant shooting, which I had looked forward to repeating in the year to come. I arrived back in El Toro on November 4; two months later I was to follow all the Marines I had shipped out to Korea during 1952.

One of the outstanding memories of the El Toro tour was the very fine relationships enjoyed with the civilian community, particularly with the Long Beach chamber of commerce, and the very influential members of Los Amigos Viejos – an exclusive social and civic order. An outstanding pillar of local support for military interests was Mike Irvine, community leader and president of the Irvine Company, our erstwhile landlord.

I recall spending an entire Saturday afternoon with Mike, exploring (in a Cadillac, no less) the ranch roads and trails of his immense domain. This entente cordial with the citizenry had not always prevailed, I was told. I still had some trouble with the local press now and then; but these misunderstandings largely evaporated after I had invited the various editors to a coffee conference in my office, which most of them had never seen.

As for the distaff operations, my wife took a very active part in Navy Relief work at El Toro, much to the satisfaction of the Station commander, Colonel Dave O'Neill, and the Washington officials of the Society. We always considered it the duty and responsibility of the general's Lady to set an example in leadership as well as in the social graces to the Ladies of the command. It was and is my sincere belief that my wife Nell excelled in both fields – thereby making my task of command infinitely easier.

Our daughter, LaVerne, now a senior at Wellesley, flew home for Christmas, through weather coming and going which caused us concern for her safety. We enjoyed a quiet family celebration, restrained perhaps by thoughts of our imminent separation for another year. Arrangements had been made for Nell to continue living in our Santa Ana quarters until she was due to leave for the East to attend LaVerne's graduation. I had hoped for some special dispensation that would permit me to attend, also; but in the event I was too busy with the last throes of the Korean War.

Shortly after the New Year had been ushered in I took leave of family and friends and took passage through the western skies to the bleak and forbidding peninsula of Korea. Once again, for the last time it proved, I was off to war.

CHAPTER XI

THE KOREAN INCIDENT

When I relieved Jerry Jerome as commander of the First Marine Air Wing in Korea, on January 8 1953, I found that I had inherited a rather far-flung empire. The wing headquarters, air control group, and one tactical fighter group (MAG 33) was based on K-3 airdrome, located on the S.E. coast of Korea at Pohang. A dive-bomber group (MAG 12) was operating from K-6, a Marine-controlled base on the West coast, south of Seoul; and a separate night fighter squadron was sharing a base with Air Force units at K-8, farther south. One helicopter squadron and a light observation squadron were based at First Marine Division headquarters in direct support of that organization; while one squadron of Corsair fighter bombers was operating from small escort carriers off the Korean West coast. At Itami A.F. Base, near Osaka in Japan, was located the Wing Service Group and the rear echelon of my Wing headquarters. Later in the year these Japanese-based units were augmented by a transport squadron, at Itami; a jet fighter group at Atsugi; and a helicopter group based initially in the Osaka area – later moved to Yokosuka.

The assigned mission of the First Marine Air Wing was to provide close air support for the Eighth Army (of which the First Marine Division was a part), air defense of the southern area of Korea, and to support the Fifth Air Force strategic bombing and deep reconnaissance missions, as required. Overall operational control of Marine air units so engaged (except for the carrier-based squadron) was vested in the Fifth Air Force command at Seoul.

This command set-up, which deprived the Marine Wing

commander of any effective operational control over his subordinate tactical units, and denied to the First Marine Division any special priority for Marine air support, had been in effect since the withdrawal of the Eighth Army from the Yalu River two years before, despite protests registered by both Marine ground and air commanders. As a matter of record my predecessors had become so reconciled to this situation that they had acquiesced in permitting the Fifth Air Force command center to issue orders direct to subordinate groups and squadrons of the Marine Air Wing, a practice which effectively bypassed the tactical command responsibility of the Marine Wing commander, reducing him to the status of a mere administrator. In truth, the Wing headquarters operational and intelligence sections had been permitted to so atrophy that the Marine Wing commander was no longer in position to exercise tactical control over his far-flung units.

A perusal of the basic command directive, which had split operational control of Marine forces in Korea between the Eight Army and Fifth Air Force, disclosed the specific proviso that "such operational control shall be exercised through the responsible commanders." Since I *was* one of the responsible commanders, the directive was clearly being violated. Armed with this cudgel I resolved to change the status quo. Obviously, the first order of business was restoration of my Wing command echelon to functional condition. I gave the necessary orders to my staff, and flew off on a series of indoctrination and inspection trips.

Midwinter among the bleak and deforested hills of Korea can be most inclement. While the temperature at Pohang seldom went below 15 degrees Fahrenheit, the constant off-sea wind at Pohang contributed a chill factor that could not be abated by the flimsy walls and oil-fired space heaters of our living quarters. Like my fictional kinsman of Robert W. Service's "Tales of the Yukon," I was always cold – until the first pink flush of the azalea shrubs on the chalky hills and the clamoring of the wild geese signaled, at long last, the advent of spring.

During these first winter months I was really too busy to worry much about physical discomfort. Despite the abominable weather I was able to keep in close and frequent touch with my subordinate commanders and their combat pilots – thanks to the instrument flying skill of my two command pilots, Major "Blanco" White, of Peruvian fame, and Captain James Gibbons. Meanwhile the reorganization of my headquarters and the routine administrative chores were being handled by my deputy, Brig. Gen. Al Kreiser, and my chief of staff, Colonel Sam Jack, able officers both, whom I had inherited from Jerry Jerome.

By mid-March I had prevailed on General Barcus, Fifth Air Force commander, a classmate from the old Air Corps Tactical School of 1937, to restore my tactical command prerogatives. About this time the somnolent ground war broke out afresh on the Marine Division front, which gave us an excuse for mounting a massive wing operation in support of the hard-pressed ground Marines. Both MAG 12 and MAG 33 were involved in this special support mission, which was led by the senior group commander, Colonel (later Major General) George Bowman. The formation was jumped by enemy MIG fighters, which were fended off with some cost to themselves, but succeeded despite this diversion in laying down a considerable tonnage of bombs and rockets at the most critical points along the Marine Division front.

Other missions of group and squadron size followed this initial stroke, much to the delight of the Marine ground troops, who had not been so favored for a long time. We had power to spare, now that I could concentrate it, so shared the wealth with the Army divisions – insofar as their sketchy ground control facilities permitted.

Once control was in my hands I had drastically reduced the number of two-and-four plane missions over the front, substituting larger formations of division and squadron size that gave better control and more effective concentration of air power. Under the more experienced flight leaders, bombing incidents involving our own troops became very rare and minor, while our losses from enemy anti-aircraft fire were materially reduced by the flak suppression tactics possible for combined fighter and dive-bomber formations. By the time Barcus was relieved in April the First Marine Air Wing was in its stride, and I encountered little difficulty in continuing the same command arrangements with the new Fifth Air Force commander, Lt. Gen. Sam Anderson.

Our night fighter squadron had gained such a reputation for shooting down enemy snoopers and interceptors that the Air Force B-29 units assigned night bombing missions in North Korea insisted on having Marine escorts. Our record of MIG kills rose appreciably, while more B-29s were able to return to base.

Our unarmed photographic squadron, equipped with the trim and speedy twinjet powered "Banshee" airplane, also served to support Air Force activities over all of North Korea to the Yalu River – and beyond. These planes could outrun and out-maneuver the MIG defenders at low altitudes, and their twin engines gave them a great safety factor. We never lost one to enemy action, although the pilots reported some hairy adventures now and then. The capacity of this 12-plane squadron was sufficient to handle on a routine basis the greater part of the reconnaissance and post-strike photo missions for the Fifth Air Force, and execute special photo-mapping runs for the Eighth Army on request, without in any way depriving the First Marine Division or the parent Marine Air Wing of their over-riding priority for photo-reconnaissance support. I considered this photographic squadron one of my most valuable units – although it never fired a gun.

We also had operating from K-3 a small ECM (electronic counter measures) squadron whose mission was to sniff out and pinpoint enemy radar installations. Most of this activity was confined to the East coast of Korea in support of the fast carrier task force (TF 77), which normally operated its tactical aircraft in that zone.

Thus the Marine Air Wing was actively engaged in supporting various elements of the Army, Navy, and Air Force, then operating in Korea, in addition to its normal mission of supporting the First Marine Division. These tasks, reputedly performed to the full satisfaction of the commanders concerned, added appreciably to the prestige of Marine Aviation, and contributed materially to inter-service cooperation and amity.

In early summer the enemy mounted a massive attack on the center of the Eighth Army front, broke through the defenses of a ROK (Republic of Korea) Army division and threatened to disrupt the entire front. We concentrated the full tactical strength of the Marine Wing for several days in support of this fluid and confused situation. The enemy thrust was eventually blunted by flank counterattacks, and the position was restored – not before causing some considerable alarm in the suddenly uncovered rear areas. My main airbase at K-3 had no ground troops in support, save for a small honor guard of Korean Marines. For a couple of days or so there appeared to be nothing between us and the enemy hordes but a hundred miles of empty terrain. We hastily organized a provisional defense battalion from our aviation ground personnel, reconnoitered and selected a defensive perimeter and prepared to give battle if need be. This was not the first time that Marines aviation personnel prepared to lay down their wrenches and pick up their rifles. I have no doubt that they would have performed creditably as Marine infantry, just as they did at Wake.

Another situation that caused me serious concern in my capacity as air defense commander of the southern area was the exposed and congested air bases on the West coast of Korea. These were sited within twenty minutes jet flying time from possible Chinese Communist air bases on the Shantung peninsula across the narrow Yellow Sea. Although theoretically I would have available on call Air Force interceptor squadrons to counter such possible attacks, I had serious doubts that they could be scrambled quickly enough for an effective defense. We could easily have suffered a major debacle, with very little warning, had the Chinese so elected.

During these months of active combat we were favored by frequent visits from the Commandant, and from the commander, Fleet Marine Force Pacific, neither of whom was in my operational chain of command, both of whom were my administrative superiors. Some of these visits had elements of "nit picking," a typical staff malady apt to be impatiently, if discreetly, resented by commanders busy with a shooting war. Lem Shepherd, in particular, often showed a deplorable lack of knowledge about aviation matters, which he was inclined to cover up by derogatory comment concerning some trivial violation of uniform regulations! He was not a popular visitor with the Marine aviators in Korea.

K-3, being the nearest air base to Japan, was used as a port of entry by a constant stream of visitors. My headquarters mess, under the competent direction of M/Sgt. Cole, became famous for the hospitality dispensed. We never lacked long for distinguished company, which on occasion included the President and Madame Rhee, Ambassador and Mrs. Ellis Briggs (who favored us most during the hunting season), high-ranking officers of all the Services, civilian officials from Washington, and other dignitaries – most of whom were old friends and acquaintances from other days, thus doubly welcome. We were never permitted to be lonely.

I was able eventually to return most of those visits, enjoying the hospitality of the Presidential Palace and American Embassy in Seoul, and the headquarters messes of my military counterparts, there and elsewhere. I also spent an occasional day afloat with my Navy friends aboard battleship or carrier, where I could observe current techniques of shore bombardment and aircraft carrier operations – a far cry from my previous experience aboard the old *Saratoga* and *Yorktown*. Life was never dull those first six months in Korea.

Mr. T. Weil, Counselor U.S. Embassy and President Syngman Rhee with Megee

I also tried to visit the front-line Marines at least monthly, timing my visits so that I might observe some air strikes being talked onto target by our sharp forward air controllers. I recall that I often visited Colonel (later General) Lew Walt, whose regiment seemed to be our most appreciative customer for close air support. (I must admit that he probably got more than his share!).

In June my good friend and ground counterpart, Maj. Gen. Al Pollock, commander of the First Marine Division, with whom I had worked so closely during these closing months of hostilities, was relieved by Maj. Gen. Ran Pate, whose tour of combat was destined to be very short. On July 27, after a last vicious effort that caused us numerous Marine casualties, the enemy adhered to the provisions of an armistice – and the guns were stilled.

Following the armistice we went into a period of consolidation and adjustment to a non-combat but combat-ready role. An intensive program of "rest and recreation" visits to Japan was established for both pilots and ground crewmen. Later we provided air transport for personnel of the First Marine Division who wished to avail themselves of the program. Training was continued, and every effort made to keep personnel occupied. Our chief worry was that post-war reaction might lead to deteriorating morale and kindred disciplinary problems. That this did not occur must be attributed to the energy and sound judgment of my subordinate commanders. They required only the light rein of guidance.

I had earlier established as one of my reforms a 100-mission normal combat tour for all tactical pilots, conforming to the rule then in effect for the Air Force. This gave the pilots something to shoot at – membership in the so-called "Century Club," which proved to be a good morale booster. Most of the officers completed their allotted stint in about six months, then did three months in non-combat assignments, or as forward air controllers. Many declined to leave their squadrons and continued to fly combat missions until departure; a very few broke under the strain and had to be relieved early. Considering the daily hazards of enemy ground fire and atrocious weather, I could only admire the fortitude of those who stuck to their cockpits, come fair weather or foul, and show compassion for their very few weaker brethren who couldn't quite make the Club. Overall, the esprit-de-

186

corps of my pilots, many of whom were reservists on temporary active duty, was superb. I gave them good group and squadron commanders, who shared their hazards in flight, encouraged them with prompt recognition of noteworthy performance, and tried to show them that my interest in their welfare and survival was not merely perfunctory. They responded with a performance beyond what a commander might reasonably expect.

One of the most moving episodes of my life occurred during the repatriation of our prisoners of war a few weeks after the armistice became effective. Arrangements had been made to exchange prisoners at a point near Panmunjom, where a suitable reception center had been constructed and staffed for the rapid processing and later evacuation of the returned prisoners. On the day scheduled for the return of the Marines who had been captured, both Ran Pate and I were in attendance. My staff had provided clean uniforms, rank insignia, and most particularly the gold wings insignia, for the first phase in the rehabilitation of our returning pilots. We met the ambulances as they came in, personally greeted each man and escorted him to the delousing unit and the showers. After they had emerged in their clean uniforms I personally pinned on each of my officers and men his insignia of rank, and on each pilot his gold wings, symbolically restoring to each his personal dignity and professional prestige. It was a most emotional moment, even for the hardiest soul among us.

Senior among the returning POWs, among the last to arrive, and quite evidently the one who had suffered the most, was Colonel Frank Schwable, former chief of staff of the Marine Wing who had been shot down more than a year earlier while flying a light transport over the front lines. Due to his rank and position his captors had subjected him to indignities and tortures beyond his capacity to endure. Certain "confessions" had thus been extorted from this highly sensitive and intelligent officer, which reflected adversely on our conduct of the war – and of course his stature as a Marine officer. He later told me that he fully expected to be met by military police and escorted to prison. When he saw me at the door of the ambulance and heard my friendly greeting he broke down completely and embraced me – later apologizing for this understandable display of emotion. Never have I seen such a dispirited and ravaged human being, one who deserved

187

from his fellow officers a full measure of compassion and understanding. That night I sent a carefully worded dispatch to General Shepherd in Washington recommending that Schwable on his return be treated with Christian charity and human compassion. I might better have saved myself the effort. The Commandant and certain of his advisors were determined to make an example of Schwable. In the end he was restored ignominiously to restricted duty where he "would never be in contact with troops." Of course he closed the case by retiring as soon as he was eligible. Even now, after twenty years, I feel that such a public crucifixion was unnecessary.

I received a number of letters later from these ex-POWs, expressing their deep appreciation for the personal reception and attention we had given them. All of them mentioned the pinning on of the gold wings as being the highlight of their rehabilitation. By far, the most articulate and profound of these letters came from Frank Schwable.

In September, the East coast of Japan was visited by a typhoon, which caused some damage to my installations and aircraft in the Tokyo-Yokosuka area. Serious flooding occurred around Osaka, leaving many Japanese villagers marooned and in danger. MAG 16 stepped into the breach with its helicopters and completed quite a massive evacuation of these people. The Japanese officials were most appreciative; the governor of the province presented me with an elaborate scroll memorializing the service of the Marine flyers to his country. Only a short eight years had passed since these same pilots were raining death and destruction on the Japanese armed forces; now we were friends and allies. *Sic transit bellum.*

The cessation of hostilities had made it possible and desirable that I devote more time to my units based in Japan. I now had the opportunity to enjoy the comforts and amenities of the adequate, well-staffed living quarters that had been provided on the Itami air base for the Marine Wing Commander. My flight log for the late summer and early fall months shows that I made very frequent trips between Korea and Japan, dividing my time between my two headquarters as the situation demanded. This helped to while away the days and weeks – and also served to keep my subordinate commanders and staff officers on their toes, had they actually needed such a stimulus!

In October we were visited separately, for the last time, by Frank Hart and Lem Shepherd, both on their bi-monthly inspection tours. After enduring these visitations I felt the need of a little "R and R" for myself. Accordingly, Rick Fairchild, my senior aide, arranged a weekend for us at a famous Japanese resort hotel, high up in the cool mountains, reserved by the American occupation forces for senior officers and their families. Being at the moment the most senior of the senior guests I was assigned to the Chrysanthemum suite, once held sacred of the Emperor. We reveled for a brief interval in sheer luxury.

My chief recreation, however, during this tour in the Far East, was found in the excellent hunting available in Korea. I had been able to get in a few fleeting excursions to the marshes and rice paddies during the spring flight of wildfowl, but after the armistice brought a more relaxed routine, I really enjoyed the fall hunting for pheasant, deer, ducks, and geese, all to be had within a half-hour jeep ride, or an hour's flight to K-6 if we really wanted geese. The deer were of the diminutive tusked variety, the geese were large and grey, of a species unknown to me, but the green-head mallards and exquisite cinnamon teal which swarmed into the rice paddies with reckless abandon were old friends. My companions on some of these hunts were Ellis and Luck Briggs, Admiral Bob Briscoe, who liked to get away from Tokyo, and the inimitable Admiral "Jocko" Clark, whose flagship, the battleship *Missouri*, often showed up off my beach with a signal asking for a helicopter. Other visitors joined in the chase now and then; some of my staff and my two aides were also avid hunters, so that our mess was never without wild game. I reveled in these autumn days, combing the brushy hills and strolling through the quaint Korean villages.

During my year in Korea I also had a contingent assignment as commander Task Force 91, which honor accrued to the senior Marine general in the Far Eastern command. This TF 91, when and if activated by the Commander Naval Forces, Far East, would comprise all Marine ground and air units in the area, and would become in effect the Fleet Marine Force, Seventh Fleet. We were never activated, of course, but since the designation gave me direct and immediate access to the senior Naval commander present (Adm. Briscoe) I found it useful on occasion.

December arrived, bringing the chill of another Korean winter. I was quite ready to turn over my command to my longtime friend, boon companion, and erstwhile flight instructor, now Maj. Gen. Al Cooley, who formally relieved me on December 5th. The change of command ceremony was attended by Ambassador Briggs and a goodly number of general and flag officers from the various commands. General Wayland awarded me the Distinguished Service Medal; President Rhee had previously decorated me with Korea's highest award. I made what I considered an appropriate "farewell address to the troops," presided over a final luncheon for my distinguished guests, and then took off with my pilot, Jim Gibbons, for Japan and home.

In Japan I reviewed the special honor guard provided for my farewell visit to the headquarters of General John Hull, then commander in chief, Far East, receiving from him a special scroll attesting to my service under United Nations command.

Thus I returned home from my last war, bearing my shield, not without honor. As I left Tokyo, I released my last official dispatch to the officers and men of the First Marine Air Wing, expressing as best as I could my feeling on leaving such a great military organization, which it would have been an honor to command at any time, a vastly greater honor to have commanded in combat. Although higher rank and greater responsibility were to come my way later, the apex of my military career was reached in Korea.

CHAPTER XII

APPROACH TO HIGH COMMAND

Pearl Harbor – 1954-1955

My new assignment had been announced in October; my orders directed me to report to Pearl Harbor as deputy commander, Fleet Marine Force, Pacific. I had hoped to get one of the two major aviation commands in the United States, but Frank hart had asked for me, particularly. After Korea, Honolulu would provide gracious living for my family and a more relaxed official routine for me. I was not too unhappy at the prospect of resting for a while on my laurels – such as they were.

I paused at Pearl Harbor but briefly, then continued my flight to Washington, where once more I was reunited with my two girls, mother and daughter. We had suffered a long absence; the hearts were indeed tender. We celebrated a joyous Christmas in our upper Connecticut Avenue apartment before packing up for the trip to Honolulu. I'd had enough trans-Pacific flying for a while so we embarked at San Francisco on an Army transport for a leisurely voyage to the islands.

Our quarters on Makalapa Drive (Admirals Row) were ready for us, and we were immediately caught up in a social whirl. The Marines at Kaneohe tendered us a welcoming reception, the aftermath of which was to profoundly affect all our lives.

Our host had told of two of his young bachelor officers to "look out for the general's daughter." Since they had never seen her I suspect that this order was received with some trepidation. One of the victims, however, took a long look at his charge and decided that he would like to make the assignment permanent. After that night, Lt. Al Broad, of Texas, was a frequent visitor in our household. He did not, I recall, lack for competition.

I had made a number of civilian friends during my wartime tour at Ewa, so that we did enjoy many social contacts among the town people. Perhaps the most valued of these friends were Bishop and Mrs. Harry Kennedy, whose Episcopal diocese extended to Japan and Korea. Since we found no Presbyterian church on Oahu, we adopted the Cathedral of Honolulu as our church home.

General Frank Hart, my immediate superior within the Marine command, and his talented wife, Katherine, made us feel socially at home within the military community; as did Admiral Felix Stump, the commander, Pacific Fleet, and his gracious young wife, Betty. We had previously served with the Stumps at Norfolk; I had known him for many years, and since he was one of the early birds in Naval Aviation we had a bond in common. Now and then he would send for me for no other discernible purpose than to reminisce about earlier adventures. I doubt that Felix Stump ever relaxed to that extent with one of his subordinate rear admirals. Felix had a reputation for irascibility.

Frank Hart also was noted for his short fuse; while Katherine was reputed to be "difficult." We never found them so; our relationships, personal and professional, were always relaxed and pleasant.

I found my duties as deputy commander quite flexible. I often represented General Hart at social and civic functions, and took my turn at visiting and inspecting our far-flung subordinate units. When Hart was absent on his many trips I also stood in for him at Pearl Harbor as "acting" commander. In short, I had a status comparable to the Vice President – ornamental, ceremonial, useful to a degree, but with no power of my own. After a few months of this lotus life I began to feel caught up in a professional back eddy.

Early in March I made my first inspection visit to our West Coast units, stopping off at El Toro, Camp Pendleton. Twenty-nine Palms, and the San Diego area. I managed to do all this within a week, which sufficed for indoctrination and personal reunions with various old friends.

Two months later I was off for the Philippines and such island way stations as Kwajalien and Guam, where we had Marine detachments serving at the various U.S. Naval Stations. This "Southern swing," as we called it, was quite a chore due to the distances involved and the humid heat of the tropics, but Frank Hart or I made the rounds quarterly. The Marine commanders whom we thus visited were of field rank (colonel and below), so that usually we were the personal guests of the Naval flag officer commanding in that particular area. The Naval commander, Philippines, who maintained his headquarters at Sangley Point on Manila Bay, was then Rear Admiral Hugh Goodwin, one of the aviation fraternity whose hospitality I always enjoyed while passing through Manila.

It was during this visit, possibly a later one, that I had the honor of being presented, at one of Hugh's official receptions, to the aged and venerable Filipino patriot, Aguinaldo, who had led the Philippine insurrection against us more than a half century before. This was dusting off some very old pages of our history. Also, on one of my visits, I had the pleasure of dining with Admiral and Mrs. Spruance, who then presided over our Embassy in Manila. It will be remembered that I served under Admiral Spruance during the War, when he was overall Fleet commander during the amphibious campaigns of Iwo Jima and Okinawa. This brief reunion with a great American and his most charming and gracious wife was for me an important milestone along the road of my experiences. Ambassador Spruance had as his counselor of embassy one of the most able of our career diplomats, the versatile and distinguished Chip Bohlen of Moscow fame. Between them I felt that never had my country been so well represented abroad.

These isolated Marine Barracks, for which we had only administrative responsibility, were at the end of the line for supply and personnel replacement, so often felt neglected – with or without reason. It was our job to dispel those feelings, adjust minor disciplinary problems where the prestige of a Marine general officer could be useful to the responsible authorities.

In June, General Hart, who had been mildly feuding with Lem Shepherd, turned into the hospital for a pre-retirement checkup. I succeeded temporarily to the command pending the arrival of his designated relief, Maj. Gen. Bob Pepper, then commanding the Third Marine Division in Japan. Bob, whom I had first known in Haiti, was some years my senior, and the logical successor to the command. He reported in on August 1st, pinned on his third star, and departed immediately on thirty days leave, thus extending my temporary command status to cover the rest of the summer. During this period we were visited by the Commandant, whom we wined and dined in appropriate fashion, and by the venerable Korean president, Syngman Rhee – for whose security I was made personally responsible by direction of Admiral Stump. In true military fashion I passed this buck on to Colonel Joe McHaney, the naval district Marine officer, whose job it should have been in the first place.

General Pepper, with his wife and two daughters, returned to Hawaii early in September, was duly feted in the best aloha tradition, and took over his command. I was now free to resume my travels.

In October I revisited Japan and Korea, including a side trip to Taiwan (Formosa) and Hong Kong. I found a changed atmosphere within the First Marine Air Wing command, and was a little saddened to note, among other erosions, a very obvious deterioration of the Wing commander's mess. The Wing had absorbed the impact of three successive commanders since my departure less than a year previously (Al Cooley had retired for physical disability after a very short tour). I was not impressed with the then-current regime. The First Marine Division merited a higher mark – perhaps I was less critical.

The visit to Taiwan was my first. I was received with full military honors in Taipei, given a tour of the Nationalist Chinese Department of Defense, quartered in a comfortable guesthouse with my staff. Our resident host was the urbane General Huang, who had been educated in the United States. He regaled us at dinner with a most enlightening and witty commentary on the ancient customs and rituals of his people. Having observed the amenities, which included an audience with the Generalissimo, President Chiang Kai-shek, we flew down the length of the big island to the Naval base on its southern tip, where a military advisory group of U.S. Marines was teaching the Chinese navy and Marine Corps the finer points of amphibious warfare. They had apt pupils. At this time there was active fighting between the two Chinese governments over the offshore islands, which lent to this training an immediate grim purpose. After witnessing the remarkable demonstration put on for my benefit by the Nationalist Chinese

Marines I doubted that Taiwan would ever be taken by the Communists.

The Chinese Admiral Liang proved to be another gracious host, who delighted in proposing interminable dinner toasts in the heady rice wine that appeared to be their national drink. He roared with laughter when I detected his subterfuge – his glass was always refilled with cold tea.

The chief U.S. Marine instructor was the personable and able young Bob Carney, son of the famous Admiral Dick Carney, wartime chief of staff 6o Admiral Halsey and now Chief of Naval Operations in Washington. Despite this paternal handicap young Bob was later to pin on stars of his own.

Vernon with Chiang Kai-shek, President of Nationalist China, 1959

195

After Taiwan we dropped into Hong Kong, which I hadn't seen since my return from China Coast in 1928. After a limited shopping spree we returned to Pearl Harbor via Manila, Guam, and Kwajalien, stopping at each of these places long enough to inspect the resident Marine units.

Meanwhile, our daughter, LaVerne, had announced her betrothal to Al Broad. We arranged for her a prenuptial visit with Al's family in Texas, after which her engagement was announced. They were married on December 21st by Bishop Kennedy in the Cathedral of Honolulu, a beautiful formal military wedding, witnessed by our many civilian and military friends. We chose the Submarine Base Officers Club for the reception (courtesy of Vice Admiral George Russell), a most gala occasion for the newly weds and the many guests – not so gala perhaps for the parents of the bride, whose enjoyment was tempered by conflict between sadness and pride in our lovely daughter and her handsome uniformed groom.

LaVerne's matron of honor was Joyce Geiger Johnson, at whose wedding in Quantico some eighteen years before LaVerne had been a flower girl. Joyce's little daughter, Melanie, now in turn was LaVerne's flower girl; thus binding the Geiger-Megee dynasties. We regretted that General Geiger had not lived to enjoy the occasion.

After our children had left for New York, where Al was to begin his civilian career with one of the leading brokerage firms, it had not seemed worthwhile to go through the empty motions of a personal Christmas celebration. Thus we were surprised and deeply touched when our aide, Captain Jack Robbins, and his petite and charming wife, Barbara, (who had also been one of LaVerne's bridesmaids), appeared on Christmas Eve with a tree and trimmings which they proceeded to install in our living room. We rallied from our momentary depression and joined in the fun, making it a truly merry Christmas. This act of thoughtful kindness endeared this young couple to us for all time to come.

Early in January I was called to Headquarters, Marine Corps, for board duty, from which I did not return until February 24, hardly a welcome absence as far as my wife was concerned. This ended my travel, pending my final detachment on June 15.

I had previously been informed by Lem Shepherd that I would be going to Washington as relief for Oscar Brice, the then-director of Marine Aviation. Since this involved a promotion to three-star rank, we were of course quite pleased. It was not to be. Frank Schilt, the senior Marine aviator, felt that he should have this job and protested so vehemently to Shepherd that my assignment was cancelled and I was ordered instead to Norfolk as commander, Aircraft, Fleet Marine Force, Atlantic. I couldn't blame Schilt for so ably defending his interests, but I did blame Lem Shepherd for the personal embarrassment his change of mind caused me.

We flew back to San Francisco in Admiral Stump's command plane, graciously lent us for the occasion, picked up a new Oldsmobile in Lansing, Mich., drove leisurely to New York for the first visit with our children, thence to Washington for temporary duty on the selection board. We did not reach our new command in Norfolk until July 20.

Norfolk – 1955

We moved again into the old "West Virginia House," which we had vacated six years before, searching the seventeen closets of that commodious establishment for any skeletons we might have left behind in 1949.

Upon reporting I received an additional duty assignment as deputy commander, Fleet Marine Force, Atlantic, under the courtly and scholarly General Oliver P. Smith. Since our separate headquarters were located in adjoining buildings I found this second hat not too binding. I seldom had to occupy the deputy commander's office provided for me, since General Smith and I settled most of our business over a daily leisurely luncheon in his mess.

Scarcely had we settled in when I was invited to accompany the Fleet commander, Admiral Jerry Wright, and his staff, to Sandia A.F. Base, near Albuquerque, for a weeklong indoctrination in atomic weapons. Then followed in September an official visit to Marine air stations at Miami and Cherry Point.

Early in October I flew to Paris, via Newfoundland, the Azores, and London, to make a presentation on amphibious warfare to the NATO War College. Later we visited Naval and Marine installations in Naples, Madrid, and Port Lyautey, Morocco, coming home via the

197

Azores and Bermuda. I particularly enjoyed Madrid, where my facility with the local language made me persona grata with the Spanish military and naval officers. We flew across Spain for an aerial view of Rota, where we later were to have base rights, then circled the Rock of Gibraltar before crossing the strait to the African littoral.

While at Port Lyautey we received a message announcing the appointment of Gen. Ran Pate, who had been serving as Assistant Commandant to Shepherd, to be the next Commandant of the Marine Corps. This was not entirely unexpected, insofar as I was concerned, knowing Shepherd's preferences, but many Marine officers were surprised at what they considered a rather pedestrian choice. We had other generals of more distinguished records who might have been selected. I had never considered myself one of the possible candidates, so could take an objective view of the matter.

Meanwhile General O. P. Smith had retired, prematurely, to make way for a friend of Lem Shepherd. While I was on excellent terms with Houston Noble, the incoming relief for Smith, I did resent the shabby treatment accorded the latter. General O.P. Smith had covered himself with immortal glory during the Korean War when he successfully extricated the First Marine Division from the frozen Chosin Reservoir area, despite the constant attacks by overwhelming hordes of Chinese Communists. Perhaps he had won too much glory, or perhaps he had simply ruffled Lem Shepherd somewhere along the way. In my opinion O.P. Smith should have succeeded to the Commandancy for the greater good of the Marine Corps – but no one asked for my opinion.

In any event I did not have long to reflect on these speculations. While we were observing a minor amphibious exercise off the beaches of Camp Lejeune, Ran Pate flew down and offered me the assignment as assistant commandant and chief of staff of the Marine Corps. I was flattered, naturally, and a little surprised. Pate, up until the time Shepherd had appointed him as assistant commandant, had always been junior to me in rank, and thus might have felt that I would prefer not to serve under him. I had no such compunctions, so I accepted the promotion with expressed appreciation, even though it involved a second move within six months.

Lem Shepherd retired on December 31, ending an era of thirty-seven years during which the Marine Corps had been largely controlled by veterans of the Fourth Marine Brigade, who had served in France during World War One. There were and still are many Marine officers who considered him a great Commandant; others had certain reservations. In any event we were pleased to stand with our contemporaries on a cold winter day and speed him on his way to the Virginia countryside. The next day Ran Pate and I each pinned on additional stars – rather, had them pinned on by our proud womenfolk – and set out to establish the new regime.

We occupied a borrowed residence in Washington (courtesy of Merwin Silverthorn, now retired and traveling) for several weeks, then moved into our newly renovated and commodious quarters at the historic old Marine Barracks at 8th and Eye streets, S.E. The three-story brick house, which was to be our home for almost two years, had been built in 1908, in an era when spacious elegance was expected in the living quarters of a senior military officer. It faced across the tree-bordered parade ground the barracks where I had lived as an officer cadet in 1921. It had taken me thirty-five years to move across that parade ground to "Quarters One."

Lieutenant General has his third star pinned on by his wife Nell and his daughter, Mrs. LaVerne Broad, at the Marine Barracks, Washington, D.C., January 1956.

CHAPTER XIII

HIGH COMMAND

Washington – 1956-1957

My appointment as Number Two in the Marine Corps hierarchy was applauded by my friends, particularly the aviators; and the press made much of the fact that I was the first Marine aviator to reach this pinnacle. I believe that the majority of the Marine officers approved General Pate's obvious gesture toward better integration of the ground and air elements of the Corps. There were, unfortunately, a few prominent dissenters among Pate's erstwhile sponsors and supporters who felt that no Marine aviator could be qualified to serve as chief of staff of the Corps. These individuals were to cause me trouble later on.

However, the transfer of command and responsibility went smoothly enough in the beginning. My indoctrination was delayed by an attack of influenza that kept me away from my office for two critical weeks and slowed me down somewhat longer period. Absence due to illness was for me such a rare event that I perhaps fretted unduly over my temporary impotence. Although this episode must have caused General Pate some inconvenience during a critical period of his own indoctrination, any qualms he may have had over my appointment were not evident to me at that time. I thought he was most considerate.

Early in the year, Oscar Brice, who had gone out to the Pacific area the past summer to relieve Bob Pepper as commander of the Fleet Marine Force, suddenly announced his intention to retire. Somewhat to my surprise, Pate offered me the job. I was sorely tempted, since this was the top field command in the Marine Corps; but decided that I couldn't very well run out on my friends and supporters who expected me to stay in Washington. I told Pate that nothing would suit me better after I had completed a normal tour in Washington – thus registering at an early date my preference for the next assignment. Of course I would be amenable to immediate transfer if for any reason he then wished to replace me in Washington. He assured me that such was not the case. He sent Al Pollock, then commandant of the Marine Corps Schools at Quantico, out to relieve Brice, and appointed Maj. General Bill Twining to relieve Pollock. This last appointment was to have some unwelcome repercussions for me during the latter part of my regime, but at the moment I thought it a logical choice.

Despite Pate's reassurances, I wondered at the time as to the real motive behind his offer to let me go after only three months in office. From what I learned later I suspect that he was already under pressure from the "dissenters" to remove me from the throne room.

Early in April the Marine Corps and the country at large suffered a traumatic experience that threatened to destroy some of our most treasured traditions. Pate was away attending a class reunion at Virginia Military Institute when I received a call from Major General Joe Berger, informing me that six recruits had been drowned at Parris Island, S.C., during an unauthorized night training march through the tidal marshes surrounding that base. The circumstances indicated gross negligence, and the implications were such that I immediately informed the Secretary of the Navy, Tom Gates. He asked that I inform General Pate that he should immediately go in person to Parris Island to forestall what would be a storm of public criticism. I dispatched the Commandant's plane to the nearest airport in Lexington, and reached him by telephone a few minutes before he was to address his reunion at VMI. This was the beginning of a hectic summer, during which Marine Corps Headquarters could do little more than try to fend off irate congressmen, outraged parents, anti-military organizations, and unfriendly press agents. Pate was sorely beset from all quarters, to the point that during the inevitable court martial of the responsible drill instructor he took certain impulsive actions that his friends considered most unwise, and which reflected most unfavorably on his judgment.

This was my first inkling of the emotional instability under pressure that was to characterize Pate's regime. Some of later actions could only be attributed to momentary mental aberrations, embarrassing though they were to his friends, and damaging to the public image of the Marine Commandant. As his principal advisor, personal confidant, and alter ego, I was caught up in a maelstrom not of my own choosing. I did what I could to shield him, and to maintain the dignity of his office, but my influence could reach only so far.

I had realized earlier that much of my time had to be devoted to matters concerning the Navy Department and the Joint Chiefs of Staff, rather than to strictly Marine Corps affairs. Consequently, I was out of my office much of the normal working day. As a result most of the routine business of the chief of staff's office had to be taken care of during the early evening hours. I seldom got home for dinner before seven o'clock. Fortunately I had a very able deputy chief of staff, Maj. Gen. Bob Hogaboom, and an outstanding staff secretary, Colonel Bob Schatzel. Aviation matters were largely delegated to Frank Schilt, still holding the chair of director of aviation and assistant commandant for air. Among us we managed to get the necessary work done.

After the Parris Island notoriety had abated somewhat, Pate began a series of long overseas trips, which kept him away from Washington much of the time. During these absences I had the additional responsibility as acting Commandant, which involved attendance at all JCS meetings, conferences with the civilian Secretaries, congressional appearances, even occasional summons to the White House. More and more the routine duties of a chief of staff devolved upon my deputy – to a greater degree than I would have preferred, since this arrangement necessarily isolated me from subordinate department heads and their daily problems. It became apparent that eventually the duties of assistant commandant and chief of staff would have to be divided between two general officers. I discussed this proposed change with Pate, but asked that it be deferred until my normal relief date.

During these two hectic years I was able to get out of Washington but infrequently and for brief intervals. I was called on for my share of public appearances, in New York, and elsewhere, which usually involved no more than an overnight absence from Washington. Quite frequently I also had to stand in for Gen. Pate on some of his scheduled speaking engagements, which otherwise would have had to be canceled by his absence. My flight log for the period records an inspection trip for the West Coast in May 1956, lasting but four days, and a shorter respite early in September to attend the National Air Races and the National Rifle Matches. Other than these occasions, my monthly flights for the year were limited to Norfolk, Camp Lejeune, and Cherry Point.

The record for 1957 shows much the same restriction on my personal movements. In May I had speaking engagements at Oklahoma State University, in Stillwater; and at the Air War College at Maxwell Field, Alabama, both alma maters. In June I made another inspection trip, coupled with speaking engagements; and in August I visited the Navy Post Graduate School at Monterrey, Calif., dropping in on the Camp Perry rifle matches on my way home. Other than these brief interludes I was chained to my desk in Washington, with but little outside diversion. Never have I had to work longer hours under so much strain for such an extended period. Quite often these hours extended right on until bedtime, as the social whirl also exacted its demands.

During the second year in Washington I was plagued with a series of problems that originated with Bill Twining and his staff at the Marine Corps Schools. Twining was a brilliant officer, and most articulate, highly regarded by Pate and Hogaboom, and certain others on my staff. He was also noted for his unfriendly attitude toward the aviation branch of the Marine Corps, and thus could hardly be included among my personal supporters.

I discovered that under the guise of instructional material the staff of MCS had published without reference to our headquarters certain doctrinal precepts that seriously restricted the employment of Marine air units in future operations. Other deviations from normal procedures made it appear that Marine Corps policies were being formulated in Quantico rather than in Washington. Since I had no intention of permitting Headquarters Marine Corps to abdicate its statutory responsibility in these matters, I had to challenge Twining. In the end Pate supported me, outwardly at least, and the objectionable material was withdrawn. Of course my actions did not endear me to Twining and his adherents – including one or two on my own staff who had to be admonished for their irregular behavior. I regretted the resultant estrangement, which has persisted through the subsequent years, but I have never regretted what I felt was my duty to do. If it comes to a choice, an officer in high places had better be respected than loved.

I returned from my June trip to the West Coast to be confronted with a most surprising *fait accompli*. During my short absence Pate had arranged for my subsequent reassignment as commander, Fleet Marine Force, Atlantic. Since I had not been consulted about this rather high-handed maneuver, I appealed directly to the Secretary of the Navy after politely informing the Commandant of my intentions. I asked for a reconsideration on the grounds that no such major command changes should be made pending the naming of a successor to Pate, whose two-year appointment would end in December. Tom Gates, the Navy Secretary, who had on occasion expressed impatience with Pate's administration, with me and directed that my transfer from Washington be deferred.

This challenge of authority, although properly accomplished, probably did not further endear me to Ran Pate and his sponsors. For my part, I felt that the personal loyalty with which I had supported Pate during his most trying months had been betrayed. I did not blame him so much as I did the "dissenters" behind the throne, who had assiduously spread the rumor that I was trying to undercut Pate and inject myself as a candidate to replace him as Commandant. This canard had no basis in fact; although due to my rank and position I was technically a candidate, I had no illusions then and few afterwards that a Marine aviator could ever be appointed to the top spot. Oddly enough, my personal relations with Pate remained outwardly cordial to the end of our association.

As autumn approached there was much speculation in Washington and elsewhere as to whether Pate would be appointed to a second two-year term. While his record of performance could hardly have been considered outstanding, the selection of a new Commandant from among the possible contenders threatened to upset the already bewildered Marine Corps – no matter where the plum fell. After due consideration and with some hesitancy, Tom Gates decided that the lesser of evils would be to reappoint Pate. While this did not turn out to be a wise decision, I must admit that at the time it seemed a wise move, an opinion elicited by the Secretary.

When this decision was announced, I collected my pound of flesh by asking for the Pacific command assignment. This involved moving Pollock prematurely, but by this time the "dissenters" were willing to take any step necessary to get me out of Washington. I was quite ready to go.

Although it was my misfortune to serve directly under a weak Commandant (compared with some of the professional giants who had previously graced that position), I do not regret the experience. High command at the seat of government has its heady compensations. For a while I was privileged to sit in the councils of the mighty.

Although my official life in Washington did not always run smoothly, my private life in Quarters One was much more tranquil. Despite official pressures, we enjoyed a pleasant and stimulating social life in the capital city, in both diplomatic and military circles. We were particularly welcomed at the Peruvian and Korean embassies, I remember, due to our previous service in those countries and our personal friendship with some of their people. We occupied a well-appointed home, well-staffed, and presided over by my lady whom I have always been pleased to consider as a gracious, charming and most competent hostess, who could and did relieve me of all responsibility for the social part of our life. We found that living on the historic old parade ground at Eighth and Eye streets, was a tranquilizing experience after the vicissitudes of official life. We often invited our friends in to share our comfort, and to be thrilled by the colorful ceremonial exhibitions at which our Marines so excelled. Of course we also participated in the official garden parties, receptions, and other entertainments given by the Pates at the adjacent Commandant's House, the oldest and one of the most distinguished residences in Washington.

In late 1956, LaVerne and her husband, Al Broad, were transferred from New York to Austin, Texas. On April 3rd, 1957, they announced the arrival of our first grandchild, Kathleen LaVerne. Nell left for Austin in a great flurry, to be gone for a month. I managed to fly down later for a brief visit. LaVerne and her infant returned the call in Washington later that summer, adding immeasurably to our happiness, in our new role as grandparents.

Early in November I took my long-postponed leave of absence, was detached by arrangement on the end of the month without having to return to Washington, and was ordered to report to my new assignment in Honolulu by December 10th. We had a delightful vacation with our children in Austin, drove across country to San Francisco, flying to Honolulu via Pan American Airlines. Among the passengers on that flight was General Claire Chennault, famous commander of the Flying Tigers in China, with his Chinese wife and small daughters. I had known him as a captain in Maxwell Field in 1937, so was pleased after twenty years to renew the acquaintance.

The welcoming party at the airport was overwhelming; we had not realized that we had so many friends in the Islands. We were whisked away to a lovely seaside suite in the Royal Hawaiian Hotel to await the renovation of our new quarters at 35 Makalapa Drive, and then proceeded to embark on what proved to be a pleasant and relaxing terminal duty assignment.

The Pacific Area – 1958-1959

We were settled in our Makalapa quarters barely in time to celebrate Christmas. Meanwhile, though, we had entertained the Commandant and his party while still living at the Royal Hawaiian. Gen. Pate had insisted on visiting the Far East at this time, which bit of obstinacy caused me some small inconvenience in that it restricted our ability to entertain him properly on such short notice.

I allowed myself two months in which to reorganize my staff and catch up on events before beginning my own official traveling. The deputy commander, Maj. Gen. Sam Jack, from whom I had received command (Pollock having departed before my arrival), left in January for his new post at El Toro. His designated relief, Maj. Gen. Francis McAllister, who had been deputy chief of staff of FMF PAC during my previous tour in Honolulu, was still commanding the Third Marine Division on Okinawa and would not be available to me until late spring. Thus I would be without a resident deputy for several weeks. I had brought Colonel Dutch Schatzel out with me to be my chief of staff, and had selected two new aides, Major Mart Williams, an aviator, and Major Jim Reeder, a ground officer. Reeder had previously served in Honolulu as aide to General Frank Hart, thus was particularly well qualified to be our social aide. Williams was to handle the military side of the office – although they both proved versatile enough to interchange duties as required.

The First Marine Force, Pacific (FMF PAC), comprised the Third Marine Division, in Okinawa (moved from Japan during my tour in Washington), the First Marine Division, at Camp Pendleton, the Third Marine Air Wing in El Toro, the First Marine Brigade at Kaneohe Bay, the Headquarters and Services Battalion at Camp Smith (Pearl Harbor), and the Force Artillery units at Twenty-nine Palms in the California desert. The two air units were grouped under the administrative headquarters of Aircraft, Fleet Marine Force, Pacific (AIR FMFF PAC) at El Toro. Even at reduced strength my command totaled more than 60,000 officers and men, two thirds of the operational strength of the entire Marine Corps. In addition, my headquarters had administrative responsibility for all Marine barracks and detachments serving the U.S. Naval bases and stations in the Pacific area.

There had been many changes in the new FMF PAC troop list, and in the location of some of the major units, during my two and one-half years absence. Consequently I was anxious to get out to the Far East, particularly, for an indoctrination and inspection trip.

Scheduled maneuvers of the Seventh Fleet, involving an amphibious operation on the west coast of Luzon seemed to offer me an excellent opportunity to see the Third Marine Division and supporting elements of the First Marine Air Wing in action. The Fleet commander, Vice Admiral Wallace Beakely, an old shipmate from the National War College and the Pentagon, invited me to observe the amphibious phase of the exercise from his flagship, the cruiser *Rochester*.

I departed from Pearl Harbor on 22 February, accompanied only by my Force engineering officer, Wally Lewis, and my aide, Mert Williams, plus the plane crew. We were then still flying the vast Pacific wastes in the ancient R5D aircraft, the four-engine, propeller driven, unpressurized work horse of World War II and the Korean incident. This aircraft did not offer too much comfort on long, over-water flights at eight or nine thousand feet of altitude, but it was sturdy and reliable. My plane had sleeping cabins and kitchen facilities (presided over on this trip by my new steward, M/Sgt. Alexander Jones), so that we did not suffer unduly en route.

We spent a day on the atoll of Kwajalien, inspecting the small Marine garrison serving that U.S. Naval base. One day in Guam sufficed for the Marine Barracks there, then commanded by Colonel Archie Vandergrift, son of our wartime Commandant, after which we flew on to Sangley Point, on Manila Bay, After a courtesy visit with the commander, U.S. Naval Forces, Philippines, then Rear Admiral Bat Cruise, one of our more salty Naval aviators, we transferred to a smaller plane for a short hop to Subic Bay and our first look at the new Naval air base which had been gouged out of the coastal mountains at prodigious labor and expense. Here I found the forward command echelon of the First Marine Air Wing and one of its component groups. Another Marine air group was operating from an interior field, courtesy of the Filipino Air Force. The Third Marine Division was then embarked in transport shipping, en route from Okinawa. After inspecting the air units I flew over to Dingalen Bay and joined Adm. Beakely on the eve of the scheduled exercises.

For this operation the Marine ground and air elements had been combined under a provisional Marine expeditionary force headquarters. As overall commander I had designated Maj. Gen. Dave Shoup, who had but recently reported to me as relief for McAllister, effective after the Third Division had returned to Okinawa. This was the first time Shoup had served under me in the field, but I knew him well personally and professionally and considered him well qualified for this temporary assignment. In any event the landing and subsequent maneuvers ashore with the Filipino Army forces went off exceptionally well, I thought, meriting the traditional "well done" signal to all commanders concerned.

The Filipino element had been particularly outstanding, in their march and camouflage discipline, and in their ability to operate effectively with a minimum of the equipment and supplies which American military men have come to feel indispensable. I thought at the time that these tough little brown soldiers would be most valuable allies on any campaign we might later have to wage in Southeast Asia.

I spent several days ashore with the troops, returned to Manila for the required protocol observations with Filipino officials, then proceeded to Okinawa for an inspection of the Third Division camps and facilities. These I found in generally miserable condition from a not-too-recent, typhoon-induced flooding of the area. I could only deplore the evident lack of initiative on the part of the responsible commanders for not sooner having dug themselves out of the mud. This sour note detracted somewhat from the excellent tactical performance demonstrated by the Third Division during the recent maneuvers. An immediate improvement in troop living conditions was noted for implementation by Shoup, as his first order of business.

My next port of call was Iwakuni, Japan, the new home of the First Marine Air Wing and two of its component groups. As I now recall, the Wing was then under command of Maj. Gen. Art Binney. The squadrons which had participated in the Philippine maneuvers were then returning to their home bases, so that I was able to complete my inspection during the week that I spent in Japan – divided between Iwakuni and Atsugi (where a former protégé of mine, Colonel Jim Mueller, was commanding Marine Air Group 13, equipped with jet fighters). Aside from the inevitable morale problems arising from a prolonged absence of family life, I found my old combat command to be in an excellent status of readiness.

My schedule included a side trip to Seoul, Korea, where I paid my respects to the President and Madame Rhee, and exchanged protocol with various commanders of the Republic of Korea (ROK) armed forces. I also called on General George Decker, then commanding the Eighth Army – still stationed along the demarcation zone after some five years of uneasy armistice.

While in Japan I also inspected the Marine Barracks at Yokosuka and Sasebo Naval Bases, and at the Naval Air Station, Atsugi, a necessary concomitant of "showing the flag" to these small Marine commands.

We returned via Midway and the gooney bird colony, arriving in Pearl Harbor after an absence of 24 days. I have recounted this trip in considerable detail as typical of my semi-annual visits to the Far East. It will be remembered that interim inspections were conducted by the deputy commander, FMF PAC, so that all units of my far-flung command received inquisitorial attention at least quarterly.

Some six weeks later I was off again to visit my West Coast units at El Toro, Camp Pendleton, and Twenty-nine Palms. This jaunt also included a cross-country flight to Washington, to attend an amphibious warfare conference, after which I enjoyed a few days leave in Oklahoma and Texas. The final stage, prior to returning to home base, involved staff visits with Navy and Marine commands in the San Diego area. I probably stayed with Al and Alice Cooley in Coronado, my usual practice.

I was back in Pearl Harbor by June 13, for a brief but vain attempt to catch up on current events before being recalled to Washington in July to attend the general officers conference at Marine Corps Headquarters. Once again I was able to steal a few hours en route to visit LaVerne and her little family.

For the next three months I settled down in my swivel chair in Pearl Harbor and really became acquainted with my own headquarters. Our old friend and mentor, Admiral Felix Stump, had been relieved as commander in chief, Pacific, by a younger contemporary of mine from carrier days, Admiral Don Felt. Another old friend, Admiral Herbert Hopwood, had taken over as commander in chief, Pacific Fleet. The entente cordial prevailed in Navy-Marine circles.

That summer the Pearl Harbor-based commands suffered a veritable plague of VIP visitors, foreign and domestic, and we were asked to do our share in their entertainment. My Lady, as usual, rose ably to the occasion with a series of formal dinners and luncheons. One of the highlights of our program was a formal luncheon that we hosted for His Imperial Majesty, the Shah of Iran, following his attendance at an amphibious exercise staged by our Kaneohe Marines. The royal gift to his hostess was an exquisitely engraved and gold inlaid silver tray – which hopefully will become a valued ancestral heirloom for future generations of the Megee dynasty.

The Austin American

A ROYAL VISIT — The Shah of Iran walks with Lt. Gen. Vernon Megee, now of Austin, in this photograph taken at the Royal Hawaiian Hotel on Thursday, June 19, 1958, prior to the Shah's visit to the First Marine Brigade at Kaneohe Bay.

Austin Couple Remembers Playing Host to Shah of Iran

Q. — What would you have said in June 1958, if the commander of United States forces in the Pacific Ocean, Admiral Felix Stump, had told you: "The Shah of Iran, that land of Omar Khayyam, is coming on an unofficial visit to Hawaii; you are in charge of His Majesty's visit?"

A. — Like four-star retired General Vernon E. Megee, then a three-star Marine Corps General, and now residing at 1122 Colorado Street in Austin, you'd have said: "Yes Sir."

His first order of business was to retrieve Mrs. Megee, who was visiting their daughter, in the continental USA. Mrs. Megee would feel at home entertaining the Shah of Iran.

And she came flying to Gen. Megee's aid.

Many centuries ago, the predecessors of this Shah had brought military chariots and armed horsemen to a peak of military perfection.

This 20th Century Shah had heard that the United States had obsoleted cavalry as a military force, and he wanted to know "how." After WW II, the Marine Corps developed a combination of moving many troops from transports to shore via double-deck landing-track vehicles. Then Marines were landed from helicopters, operating from aircraft carriers, on their flanks and

The Austin American: Austin Couple Remembers Playing Host to Shah of Iran

211

After my November trip to the Far East I took along my chief of staff and his four principal assistants. was our aide and general tour manager, while M/Sgt Charley Meyers saw that we were properly fed en route. We spent three days in the Tokyo-Atsugi area, overnighted at Iwakuni before flying on to Seoul for a fast round of amenities. From there our flight route took us over my old headquarters base at K-3, now occupied by the Korean Air Force, across the Strait of Korea and the Eastern Sea directly to the main island of Okinawa. Here we concentrated our efforts on the Third Marine Division for four full days of staff visits and unit inspections. We found the physical condition of the camps much improved, but there were apparent certain undercurrents of poor morale which disturbed me. Shoup's method of handling his officers did not appeal to me – but I managed to keep my own counsel for the moment at least.

This trip included two days on Taiwan (Formosa) to observe scheduled maneuvers of our allied Chinese armed forces, followed by the usual protocol visits in Taipei, the capital city, attended by overmuch oriental banqueting. Following these gastronomic marathons we felt the need of some "rest and relaxation," so flew over to Hong Kong for an overnight stay. I might mention that I was favorably impressed with the training of the Nationalist Chinese armed forces – particularly their Marines and jet fighter pilots. It appeared to me that they were very staunch allies and worthy of our support.

Returning, we spent the night on Okinawa, before departing homeward, via Japan.

We checked in at Pearl Harbor on 2 December. My wife had departed for Texas meanwhile for a Christmas visit with our children. I spent a brief ten days in my office before taking off again for the West Coast and Washington, with ten days delay in returning authorized. We enjoyed a most pleasant family Christmas together in Austin. I was back in Pearl Harbor by the end of the month. It had been a busy year.

Nell arrived a few days later, bringing with her for an extended visit our daughter and granddaughter. Before they returned to Austin in the spring the two-year-old Kathleen had been utterly spoiled, I fear, by all the special attentions bestowed on "the little princess" by our entire household. In the usual fashion of grandparents we left that worry to her mother.

Early in May I was off again for the West Coast and Washington, stopping at Austin en route, as usual. While on the West Coast I spent nine days aboard the amphibious command ship, *U.S.S. Clymer*, and five days under canvas on the Camp Pendleton reservation, in connection with joint amphibious exercises involving the First Marine Division and the Third Marine Air Wing plus supporting Force units. During these maneuvers I functioned as Landing Force commander; my opposite number on the Navy side, Vice Admiral John Sylvester, was commander, Amphibious Force. This was the first and only opportunity I had during my regime to actually function at sea and ashore as a major tactical commander. I found the experience very stimulating. June 7 found me back at my desk.

Only a week later I again headed for the Far East, scheduled to visit my units in Japan and Okinawa, then to continue on a quasi-diplomatic tour of Malaya, Thailand, and Vietnam. I took with me, as the designated relief for Shoup, Maj. Gen. Bob Luckey. Art Binney had already been relieved as commander of the First Marine Air Wing by Maj. Gen. Carson Roberts, an old flying mate from the pre-war "circus squadron." In each case I considered the change in commanders beneficial for the units concerned.

Since trouble was threatening, even then, in the separate states of old Indochina, I deferred the formal change in command for the Third Marine Division so that I might take Luckey with me to Manila, Singapore, Bangkok, and Saipan. I felt that he needed to know something about the area in which his new command might have to operate.

I had never gone beyond Manila in this direction, so found the trip extremely interesting. Singapore was of course still very much colonial British, in atmosphere and customs, if not in government. Saigon was a miniature Paris with exotic trimmings, but Bangkok was strictly an oriental jewel – the city of fantastic temples. We were duly wined and dined by our resident confreres and by the foreign officials concerned. After conferring with the American Ambassador I arranged a clandestine flight to Laos for Luckey and my operations officer, Colonel (later Major General) Bob Owen. I was not permitted to join them.

We dropped off Luckey in Okinawa, and took aboard Shoup for

the return journey via Japan and Midway. We arrived back at Pearl Harbor just in time to celebrate the Fourth of July.

Before the month ended I was called back to the West Coast to attend the Fleet commanders' conference in Monterrey. This trip was extended to permit a visit to the National Rifle Matches at Camp Perry for Maj. Gen. Ed Snedeker, commander of the First Marine Division, and myself. We felt it appropriate that we personally support our competing Marine teams. I also had some important business with the officials of the National Rifle Association.

During the preceding weeks the usual intrigues and maneuvering for position attendant upon the appointment of a new Commandant were proceeding apace in Washington. As the senior field commander and the ranking officer on the permanent list I would automatically have to be considered as a leading candidate. I had generally held myself aloof from active campaigning, although I was not without support, political and otherwise. I had good reason to stand on my record and to feel that if military merit and experience were to be determining factors I might well stand a good chance of being selected. On the other hand, if Service politics were to determine the issue, the anti-aviation "dissenters" would move heaven and earth to thwart my nomination, or the nomination of any other aviator, for that matter. Being a realist, I could only too well count the odds against me.

I had reason to believe that Pate would support his old classmate, Al Pollock; Shepherd and his adherents would try to influence the vote for Bill Twining. I did not consider that there might be other serious contenders. As the summer waned the pressures on the Secretary of the Navy, Bill Franke, were intensified. In the end he surprised us all, with a Draconian decision, which passed over nine senior general officers in favor of Major General David Shoup, the Medal of Honor winner from Tarawa. Perhaps Bill Franke and his advisors felt that they needed a hero to restore the Marine Corps to glory after the Pate regime.

The Secretary conveyed his decision to me by means of a personal letter, hand delivered to me while at Camp Perry. He assured me of his continued high regard, told me that I had been seriously in the running up to the moment of final decision, and expressed the wish that I remain on active duty in support of the new administration. I appreciated the personal consideration that prompted this letter, but I felt that my usefulness to the Marine Corps had ended. I had served two separate tours, totaling four years, in the temporary rank of lieutenant general. I could not hope for further continuance in a high command position after Shoup, my former subordinate, should have assumed office. I felt no desire to revert to my permanent rank of major general and a lesser assignment for the two and one half years I had yet to go for my retirement; nor did I wish to subject myself annually to the vagaries of a "plucking board," to which because of my age and seniority I would be particularly vulnerable. Furthermore, I thought that the kindest thing I could do for Shoup was to clear the deck for any appointments he might wish to make. There was another consideration. Should I retire not later than November 1st, I would still be eligible for promotion to four-star rank on retirement, by virtue of the three combat decorations that I wore. All these factors I pointed out to Secretary Franke in a hand-written letter that I composed while riding back to the West Coast. On August 22 I submitted my formal request for retirement after "forty years of service."

The other four lieutenant generals immediately followed suit, each for personal reasons of his own. There was no collusion among us, and none made public his reasons for bowing out, insofar as I know. The press interpreted these retirements as a negative vote of confidence for the nominee, and flaunted headlines reporting that, "The Marine Corps High Command is Falling Apart." The interpretation was not without merit; the report had no basis in fact. In any event Shoup was given a clear field of fire.

For my final inspection and protocol visit to the Far East I was authorized to have my wife accompany me in my official airplane. My flight crew made the necessary alterations in the cabin arrangements to ensure her comfort on the long flights, and our faithful retainer, Charley Meyers, went along as flight steward. Major Reeder served as social aide and tour director. There were no other passengers. Our itinerary included Wake Island, Japan, Korea, Taiwan, Okinawa, and Hong Kong, returning via Japan and Midway. Our reception everywhere along this route befitted royalty – this was the first time that the Marines of the western Pacific had been permitted to see and visit with *their* first Lady. They outdid themselves in extending courtesies, as was the case for the foreign officials. In Japan we drafted my Far East Representative, Colonel Jake Goldberg, and his agreeable wife, Ann, to accompany us for the rest of the tour. Ann rose to the occasion and established herself as the "lady in waiting." We had a lot of fun together.

Perhaps the highlight of the trip for Nell was her gracious reception in Korea. We were entertained in the Palace by the President and his charming consort, Madame Rhee, provided with a sumptuous apartment with beautifully trained Korean attendants, regaled with exotic entertainment in the Korean manner. As a final touch, Nell was presented with a fitted costume dress of traditional Korean pattern, which she insisted on wearing to one of the social functions – to the unconcealed delight of the Korean ladies.

Later, Nell was to enjoy the homage paid her by the Marine officers of the Third Division and First Wing, who staged dress receptions in her honor. To top off these honors we then gave her a carte blanche shopping tour in Hong Kong, which had to be sandwiched in between a luncheon at Government House, and a full-dress dinner aboard the flagship of the British commander in chief, *Far East*, Admiral Sir Eric Gladstone, and his Lady, and I could only bask in reflected glory; this was strictly Nell's tour.

For my part I had to be content with farewell reviews and parades, each of which occasions gave me the opportunity to express my sentiments of pride and affection for those stalwart and loyal Marines who had served me so well for so long.

We returned to Honolulu drained of emotion but full of pride and happiness. There were but two weeks remaining of active duty, and there was packing to be done. The grand finale was to be my retirement ceremony staged by the First Marine Brigade at Kaneohe, attended by the great and near-great of Honolulu, from the government down. The Marines paraded in a downpour, proudly and defiantly, I thought. I read my retirement orders, added a few words of farewell. Nell and Admiral Hopwood pinned on my fourth star, the guns fired seventeen times and my new flag was run up for the ruffles and flourishes. Thus ended for me my active duty association with the Marine Corps, which had become such a vital part of my life.

The last gesture tendered us by our loyal Marines was a full-dress reception at the officers club in Kaneohe, arranged by our very special friends, Avery and Polly Kier. From there we were driven directly to Hickham Field to board our old DC-5 for the final journey to San Francisco. Choked by fragrant leis and emotion we said our farewells, and closed the door in Honolulu. Once aloft we had another pleasant surprise. Lieut. Col. McRay, our first aide, had flown out from El Toro for the occasion. He and now unveiled the final treat – a champagne steak dinner. Our cup was indeed full.

At San Francisco we said goodbye to our staff and plane crew, lingered just long enough to administer the oath of office to our successor, Tom Wornham, as our final official act. We recovered our automobile, previously shipped from Honolulu, and headed out as civilians on an extended and indefinite vacation. An era had ended.

EPILOGUE

Nineteen years have now passed since my retirement from active military duty. I have been occupied with a period of academic study at the University of Texas, leading to a master's degree. I have prepared a family genealogy, and have written several magazine articles. These memoirs have been written over a ten-year period, in between other literary tasks. My major effort, however, has been expended in the organization and direction of the Marine Military Academy, at Harlingen, Texas, where I have been superintendent, president, and finally chairman of the board. All these activities have served to ease the sometimes-difficult transition from military to civilian status.

We have traveled rather widely since retirement, an extended tour of Europe and the British Isles, several long motor trips to various parts of the States and Canada. We have made many new friends, and our social life has been adequate. Above all, however, our unaccustomed stability of residence has permitted us to participate in the formative years of our two grandchildren, the second of which, Tyson Megee Broad, was born January 3, 1962. His arrival was announced by his father, Al, in a rather laconic but highly appropriate message: "General, we have us a new hunting partner." And so we have. His training in outdoor lore and in marksmanship has by general consent devolved largely on his maternal and martial grandfather – who has been a demanding taskmaster. We have taken Tyson on two of our long motor tours, that he might profit from a first-hand study of American geography and history. No one has really seen America until he has done so in the company of a lively and inquiring ten-year-old boy; no one has ever enjoyed such a tour as much as did we, his proud grandparents.

Our Kathleen is now a lovely and willowy teenager with a flair for horses. She also loves the outdoors and is adept with a rifle and fishing rod – as becomes the granddaughter and daughter of a Marine. We are indeed blessed with handsome, healthy and intelligent grandchildren; and it has been a great comfort to have our daughter and her family as participants in our daily lives.

In final retrospect, we have enjoyed a very full and satisfying life, even exotic and glamorous at times. Our military career spanned a period of exceptional development for the Marine Corps; we are proud to have been a part of it. The memories of our forty years of active service, and of those boon companions of yesterday, serve to sustain us during our sunset years.

The *Navy Times*, May 17, 1958

Lt. Gen. Vernon E. Megee, U.S.M.C.
Commanding General Fleet Marine Force, Pacific

Vernon E. "Maggie" Megee rose from Marine Private to Three-Star General to command two-thirds of the operating forces of the Marine Corps – the Fleet Marine Force, Pacific.

The General was born June 5, 1900, at Tulsa, Okla., and is a graduate of Oklahoma A&M College. He enlisted in the marines in 1919 and was commissioned a second lieutenant three years later.

His first 10 years in the corps were spent as a ground marine. He's "crunched more gravel" than many of the ground officers on his staff. As an aviator, he played a major role in developing the marine doctrine of close air support for ground forces.

During the marine assault on Iwo Jima during W.W.II, a flight of 48 Corsairs and Hellcats blanketed the front and flanks along the beaches, putting their fire 200 yards ahead of the advancing troops. The plan for this "truly close" air support was drawn up by then-Col. Megee, who told his pilots to "go in and scrape your bellies on the sand." They did – then and many times later.

Gen. Megee follows his own dictum in his present job. He flies far and "scrapes his belly on the sand" to find out first hand what his troops are doing and what needs to be done.

The General's Scotch-Irish and French Huguenot ancestry is reflected in his hammering insistence on military fundamentals and military frugality. It also shows up in his stiff-backed code of personal ethics and in his short temper. He often writes letters twice – one to be torn up after he's vented his feelings, the other to be signed and mailed. Neither leaves any doubt as to his views.

An Oklahoma ranch background has given Gen. Megee a keen interest in fishing, hunting, and gun collecting, and an interest in marksmanship which is exceptionally high, even for a marine. He's a familiar figure on the firing line wherever rifle and pistol matches are held, and is a formidable opponent on the skeet range.

Just before his present assignment, Gen. Megee served for 2 years as Assistant Commandant and Chief of Staff at Marine Corps Headquarters. He was a popular figure in Spanish-speaking circles in Washington because of his fluent knowledge of Spanish and South

American history and customs which he picked up while on duty in Peru some years before. He's married to the former Nell Neimeyer of Chandler, Okla. Their daughter, Nell LaVerne, is the wife of Alfred T. Broad, a Marine Reserve Officer.

Vol. 138   WASHINGTON, TUESDAY, FEBRUARY 25, 1992   No. 23

Congressional Record

United States
of America

PROCEEDINGS AND DEBATES OF THE 102^{d} CONGRESS, SECOND SESSION

United States
Government
Printing Office
SUPERINTENDENT
OF DOCUMENTS
Washington, DC 20402

OFFICIAL BUSINESS
Penalty for private use, $300

Congressional Record

E 406 CONGRESSIONAL RECORD — *Extensions of Remarks* February 25, 1992

AVIATOR VERNON MEGEE—FROM PRIVATE TO FOUR STAR GENERAL—DIES AT 91

HON. J.J. PICKLE

OF TEXAS

IN THE HOUSE OF REPRESENTATIVES

Tuesday, February 25, 1992

Mr. PICKLE. Mr. Speaker, when the history of the development of warfare of the 20th century is written by those historians who come after us, that chapter devoted to the evolution of aeronautical warfare must address the development of the importance of close air support of ground troops for the success of any military operation.

One man is credited with directing the fulfillment of that concept. Recently, that man, Vernon E. Megee, was buried in Arlington Cemetery among his fellow patriots.

The product of a mid-America one-room schoolhouse, Vernon enlisted in the Marine Corps in 1919 to earn enough money to finish his education at Oklahoma A&M. Forty years later, he retired with four stars, the only man in the Marine Corps to go from private to four-star general.

My friendship with Vernon Megee began during his retirement years in Austin from 1960 to 1989. General Megee was the ranking military officer in the area and participated in many official events, as well as being an active member of the Downtown Rotary and the Austin Country Club.

I was privileged to know this outstanding American and am proud to offer his obituary for all to read:

General Vernon E. Megee, USMC (Retired), died January 14, 1992, in Albuquerque, New Mexico, at the age of 91. General Megee, as the Colonel in command of the Landing Force Air Support Control Unit One at Iwo Jima, told his pilots to "Go in and scrape your bellies on the beach" in support of the ground troops. At the battle for Okinawa, both Marine and Army units utilized close air support under Colonel Megee's command to help "dig the enemy out of caves" as the ground units advanced. For his outstanding performances at Iwo Jima and Okinawa, General Megee was awarded the Legion of Merit with Combat "V" and the Bronze Star with Combat "V".

In 1966, General Megee became the first Marine Aviator to hold the post of Assistant Commandant/Chief of Staff of the U.S. Marine Corps. Previously, in 1950, General Megee served as the Director of Intelligence for the Joint Chiefs of Staff. He then took command of the First Marine Air Wing in Korea. His last assignment was as Commanding General, Fleet Marine Force, Pacific, where two-thirds of the combat forces of the Marine Corps were under his command. He retired from the Corps in 1959, having risen from private to four stars after more than forty years of service.

General Megee also saw foreign service in Haiti, Nicaragua, China and Peru. His other decorations included the Navy-Marine Corps medal (Nicaragua), the Cruz de Aviacion (Peru), the Military Order Taikuk with Silver Star (Korea) and the Distinguished Service Medal.

A native of Oklahoma, General Megee received a Bachelor of Science degree from Oklahoma State University and a Masters of Arts from the University of Texas.

After retirement from active duty, much of General Megee's time was spent in volunteer service to the Marine Military Academy in Harlingen, Texas, where he served as the first Superintendent and as President of the Board of Trustees. On November 11, 1988, General Megee was elevated to the position of Emeritus Chairman of the Board, the first trustee of the school to be so honored.

General Megee is survived by his daughter and son-in-law, LaVerne M. and Alfred T. Broad of Albuquerque, NM, a granddaughter, Kathleen L. Broad, also of Albuquerque, and a grandson, Tyson Megee Broad of Portland, Oregon. Two sisters, Opal Jones of Fresno, California and Walsa Meier of Broken Arrow, Oklahoma also survive General Megee. His wife, Nell, preceded him in death in July, 1989.

A SPECIAL SALUTE TO GEORGE FRASER

HON. LOUIS STOKES

OF OHIO

IN THE HOUSE OF REPRESENTATIVES

Tuesday, February 25, 1992

Mr. STOKES. Mr. Speaker, the Black Professionals Association charitable foundation will host its twelfth annual scholarship and awards gala on Saturday, February 29, 1992. The gala will be held at the Stouffer Tower City Plaza Hotel in Cleveland.

The Black Professionals Association [BPA] is composed of more than 100 black professionals throughout the Cleveland area. Over the years, BPA has chosen February, which is officially celebrated as Black History Month, to recognize African-Americans who are positive role models and have achieved significant success in their chosen fields. Last year, I had the honor of being selected as the 1991 black professional of the year by this distinguished organization.

Today, I am proud to rise to salute the 1992 black professional of the year, George C. Fraser. I would like to share with my colleagues and the Nation some of the achievements of this year's award recipient.

Mr. Speaker, George Fraser is an outstanding businessman who is the founder and president of SuccessSource, Inc. Prior to the inception of SuccessSource, Mr. Fraser was employed by the Ford Motor Co. in its Minority Dealer Development Program. In addition, he was employed by United Way Services of Cleveland and Procter & Gamble in Cleveland and Cincinnati, OH.

It was during this time, Mr. Fraser recognized the need for an informational resource which would encompass the enormous breadth and diversity of African-American excellence. His idea led to the development of the innovative SuccessGuide, a comprehensive directory of African-American businesses, professionals and organizations.

The SuccessGuide has proven to be a valuable asset to the African-American community. More importantly, by tapping into this arena, George Fraser has succeeded in overcoming the corporate barriers which in the past impacted African-Americans' ability to start and build businesses.

As president of this highly successful venture, Mr. Fraser was able to make the transition from employee to employer in only 5 years. This is a testament to this individual's persistence, his entrepreneurial skills and his determination to succeed.

Mr. Speaker, not only is George Fraser a successful businessman, but he is also a community leader. He gives his time and advice to benefit community organizations. Mr. Fraser is an active board member of the Greater Cleveland Growth Association, John Carroll University and the Cleveland NAACP. The black Professionals Association is just one of the many organizations to recognize Mr. Fraser's exceptional talents. Recently, he was selected as the role model of the year by the Teen Father Program; national volunteer of the year by the United Negro College Fund [UNCF]; and Cleveland business advocate of the year by the city of Cleveland.

Mr. Speaker, I am pleased to congratulate George Fraser for his achievements. He is well deserving of the honor accorded him as the 1992 black professional of the year. I join the community and his many friends and colleagues in saluting him on this momentous occasion, and I wish him much continued success.

TAX FAIRNESS IN THE 1980'S

HON. JOHN EDWARD PORTER

OF ILLINOIS

IN THE HOUSE OF REPRESENTATIVES

Tuesday, February 25, 1992

Mr. PORTER. Mr. Speaker, as debate within the Congress returns to the issue of tax fairness, I commend to my colleagues the following article which appeared in the Wall Street Journal on January 28, 1992. I think the author makes a convincing argument against the claim that the rich were the only ones who improved their standing during the 1980's.

[From the Wall Street Journal, Jan. 28, 1992]

THE "FORTUNE FIFTH" FALLACY

(By Richard B. McKenzie)

According to numerous pundits, the shift in the income distribution during the 1980s was seismic, with the rich getting richer and the rest, poorer. And Harvard University Prof. Robert Reich, among a chorus of academics, professes that only the "most fortunate fifth" of Americans—those with "princely incomes" (or households with more than $55,205 in annual earnings in 1990)—improved their economic status during the past two decades. He argues that justice and economy demand that the growing hardship of the lower four-fifths of income earners be relieved with whopping tax increases on the rich.

Myriad versions of these claims have often been fortified with citations of official data on real median family income and on share of income going to each of the five quintiles of households. Real median family income, adjusted for inflation using the standard consumer price index (CPI), and are relative to the 1970 level, did trend downward from 1970 to the late 1980s.

A DEFECTIVE MEASURE

Fortunately, the reality of the changing income distribution is far more complicated than the modern prophets of gloom would have us believe. As rarely conceded, the real median family income began to rebound after 1982. Moreover, this measure of the real median is defective in three key ways: (1) the method for computing the CPI was changed in 1983, the effect of which was to obscure the growth in real income; (2) the average family size fell by 17% between 1970 and 1986; and (3) fringe benefits and other wage supplements, which are not counted as family income, expanded from 12% of total wages and salaries in 1970 to 20% in 1986.

Arlington Cemetery, Feb 25, 1992

Aviator Vernon Megee – From Private
To Four Star General – Dies At 91

Hon. J.J. Pickle
Of Texas
In the House of Representatives
Tuesday, February 25, 1992

Mr. Pickle: Mr. Speaker, when the history of the development of warfare of the 20th century is written by those historians who come after us, that chapter devoted to the evolution of aeronautical warfare must address the development of the importance of close air support of ground troops for the success of any military operation.

One man is credited with directing the fulfillment of that concept. Recently, that man, Vernon E. Megee, was buried in Arlington Cemetery among his fellow patriots.

The product of a mid-America one-room schoolhouse, Vernon enlisted in the Marine Corps in 1919 to earn enough money to finish

225

his education at Oklahoma A&M. Forty years later, he retired with four stars, the only man in the Marine Corps to go from private to four-star general.

My friendship with Vernon Megee began during his retirement years in Austin from 1960 to 1989. General Megee was the ranking military officer in the area and participated in many official events, as well as being an active member of the Downtown Rotary and the Austin Country Club.

I was privileged to know this outstanding American and am proud to offer his obituary for all to read:

General Vernon E. Megee, USMC (Retired), died January 1992, in Albuquerque, New Mexico, at the age of 91. General Megee, as the Colonel in command of the Landing Force Air Support Control Unit One at Iwo Jima, told his pilots to "Go in and scrape your bellies on the beach" in support of the ground troops. At the battle for Okinawa, both Marine and Army units utilized close air support under Colonel Megee's command to help "dig the enemy out of caves" as the ground units advanced. For his outstanding performances at Iwo Jima and Okinawa, General Megee was awarded the Legion of Merit with Combat "V" and the Bronze Star with Combat "V."

In 1956, General Megee became the first Marine Aviator to hold the post of Assistant Commandant/Chief of Staff of the U.S. Marine Corps. Previously, in 1950, General Megee served as the Director of Intelligence for the Joint Chiefs of Staff. He then took command of the First Marine Air Wing in Korea. His last assignment was as Commanding General, Fleet Marine Force, Pacific, where two-thirds of the combat forces of the Marine Corps were under his command. He retired from the Corps in 1959, having risen from private to four stars after more that forty years of service.

General Megee also saw foreign service in Haiti, Nicaragua, China and Peru. His other decorations included the Navy-Marine Corps medal (Nicaragua), the Cruz de Aviacion (Peru), the Military Order Taikuk with Silver Star (Korea) and the Distinguished Service Medal.

A native of Oklahoma, General Megee received a Bachelor of Science degree from Oklahoma State University and a Masters of Arts from the University of Texas.

After retirement from active duty, much of General Megee's time was spent in volunteer service to the Marine Military Academy in Harlingen, Texas, where he served as the first Superintendent and as President of the Board of Trustees. On November 11, 1988, General Megee was elevated to the position of Emeritus Chairman of the Board, the first trustee of the school to be so honored.

General Megee is survived by his daughter and son-in-law, LaVerne M. and Alfred T. Broad of Albuquerque, NM, a granddaughter, Kathleen L. Broad, also of Albuquerque, and a grandson, Tyson Megee Broad of Portland, Oregon. Two sisters, Opal Jones of Fresno, California and Walsa Meier of Broken Arrow, Oklahoma also survive General Megee. His wife, Nell, preceded him in death in July 1989.

Marine Band, Arlington Cemetery, Feb 25, 1992

ROBERT SHERROD
4000 CATHEDRAL AVE., N.W.
WASHINGTON, D.C. 20016

202/338-7381

21 January, 1992

Dear Mrs. Broad -

My heartfelt condolences on the passing of your distinguished father. I had not seen him for many years but I remember him well during the 1948-1952 period, when I was writing my _History of Marine Corps Aviation in World War II_, for which I interviewed him many times, as the book evidences. I also had met him briefly during that war, when I was a correspondent in the Pacific for _Time_ and _Life_.

I admired "Maggie" immensely, and so did his other contemporaries. He truly was a historical figure as well as a real officer and a gentleman — and a fine companion.

Most sincerely,
Robert Sherrod

Condolence letter to LaVerne Broad from journalist Robert Sherrod

228

Text of Sherrod Letter:

January 21, 1992

Dear Mrs. Broad,

My heartfelt condolences on the passing of your distinguished father. I had not seen him for many years but I remember him well during the 1958-1952 period, when I was writing my *History of Marine Corps Aviation in World War II,* for which I interviewed him many times, as the book evidences.

I also had met him briefly during that war, when I was a correspondent in the Pacific for *Time* and *Life.*

I admired "Maggie" immensely, and so did his other contemporaries. He truly was a historical figure as well as real officer and a gentleman – and a fine companion.

Most sincerely,

Robert Sherrod

Memoirs Of A Marine

exists between th
of the United Sta

"Whereas the
war against the
be it resolved)
America in Cong
the Imperial Jc
States is here'
directed to co
the resources
government ;
resources of

It passed
referred to
passed the I
Wallace and
dissenting
against war

Extent
and enlis'
war.

STATE DE'

had call
nation
mutual
to the
the re
offici
not w

#1
the J

f
thi
nat

en
of
1

DEPARTMENT OF STATE RADIO BULLETIN
NO. 291 DECEMBER 8 1941.

This digest has been compiled from press and other sources
expression of official opinion.

WHITE HOUSE - PRESIDENT'S MESSAGE TO CONGRESS. At approximately 12:30 PM EST today
the President delivered to the Congress assembled the following
message: "Yesterday December 7 1941 - a date which will live in infamy - the
United States of America was suddenly and deliberately attacked by naval and air
forces of the Empire of Japan.

"The United States was at peace with that nation and at the solicitation of
Japan was still in conversation with it's government and it's Emperor look-
towards the maintenance of peace in the Pacific - Indeed one hour after Japanese
air squadrons had commenced bombing in Oahu the Japanese Ambassador to the United
States and his colleague delivered to the Secretary of State a formal reply to a
recent American message - while this reply stated that it seemed useless to
continue the existing diplomatic negotiations it contained no threat or hint of
war or armed attack.

"It will be recorded that the distance of Hawaii from Japan makes it obvious
that the attack was deliberately planned many days or even weeks ago - during the
intervening time the Japanese government has deliberately sought to deceive the
United States by false statements and expressions of hope for continued peace.

"The attack yesterday on the Hawaiian Islands has caused severe damage to
American naval and military forces. Very many American lives have been lost.
In addition American ships have been reported torpedoed on the high seas between
San Francisco and Honolulu.

"Yesterday Japanese government also launched an attack against Malaya.

"Last night Japanese forces attacked Hong Kong.

"Last night Japanese forces attacked Guam.

"Last night Japanese forces attacked the Philippine Islands.

"This morning the Japanese attacked Wake Island.

"Japan has therefore undertaken a surprise offensive extending throughout the
Pacific area. The acts of yesterday speak for themselves. The people of the
United States have already formed their opinions and well understand the
implications to the very life and safety of our nation.

"As Commander in Chief of the Army and Navy I have directed that all measures
be taken for our defense.

"Always will we remember the character of the onslaught against us.

"No matter how long it may take us to overcome this premeditated invasion the
American people in their righteous might will win through to absolute victory.

"I believe I interpret the will of the Congress and of the people when I
assert that we will not only defend ourselves to the uttermost but will make very
certain that this form of treachery shall never endanger us again.

"Hostilities exist. There is no blinking at the fact that our people, our
territory and our interests are in grave danger.

"With confidence in our armed forces - with the unbounding determination of
our people - we will gain the inevitable triumph - so help us God.

"I ask the Congress declare that since the unprovoked and dastardly attack by
Japan on Sunday December 7th a state of war has existed between the United States
and the Japanese Empire".

CONGRESS - WAR RESOLUTION. Following the President's message Sen Connally
introduced the following resolution: "Declaring that a state of war"

Department of State Radio Bulletin
No. 291 December 8 1941

DEPARTMENT OF STATE RADIO BULLETIN
NO. 291 DECEMBER 8 1941.

This digest has been compiled from press and other sources and is in no way an expression of official opinion.

WHITE HOUSE – PRESIDENT'S MESSAGE TO CONGRESS. At approximately 12:30 PM EST today the President delivered to the Congress assembled the following message:

"Yesterday, December 7, 1941—a date which will live in infamy—the United States of America was suddenly and deliberately attacked by naval and air forces of the empire of Japan.

"The United States was at peace with that nation and, at the solicitation of Japan, was still in conversation with its government and its emperor looking toward the maintenance of peace in the Pacific. Indeed, one hour after Japanese air squadrons had commenced bombing in Oahu, the Japanese ambassador to the United States and his colleague delivered to the secretary of state a formal reply to a recent American message. While this reply stated that it seemed useless to continue the existing diplomatic negotiations, it contained no threat or hint of war or armed attack.

"It will be recorded that the distance of Hawaii from Japan makes it obvious that the attack was deliberately planned many days or even weeks ago. During the intervening time, the Japanese government has deliberately sought to deceive the United States by false statements and expressions of hope for continued peace.

"The attack yesterday won the Hawaiian Islands has caused severe damage to American naval and military forces. Very many American lives have been lost. In addition, American ships have been reported torpedoed on the high seas between San Francisco and Honolulu.

"Yesterday Japanese government also launched an attack against Malaya.

"Last night Japanese forces attacked Hong Kong.

"Last night Japanese forces attacked Guam

"Last night the Japanese attacked Wake Island.

"This morning the Japanese attacked Midway Island.

"Japan has therefore undertaken a surprise offensive extending

231

throughout the Pacific area. The acts of yesterday speak for themselves. The people of the United States have already formed their opinions and well understand the implications to the very life and safety of our nation.

"As commander in chief of the Army and Navy, I have directed that all measures be taken for our defense.

"Always will we remember the character of the onslaught against us.

"No matter how long it may take to overcome this premeditated invasion, the American people in their righteous might will win through to absolute victory.

"I believe I interpret the will of the Congress and of the people when I assert that we will not only defend ourselves to the uttermost but will make certain that this form of treachery shall never endanger us again.

"Hostilities exist. There is no blinking at the fact that our people, our territory, and our interests are in grave danger.

"With confidence in our armed forces—with the unbounded determination of our people—we will gain the inevitable triumph, so help us God.

"I ask that the Congress declare that since the unprovoked and dastardly attack by Japan on Sunday, December 7, 1941, a state of war has existed between the United States and the Japanese Empire."

CONGRESS – WAR RESOLUTION. Following the President's message Sen. Connelly introduced the following resolution:

"Declaring that a state of war exists between the Imperial Government of Japan and the Government and the people of the United States and making provisions to prosecute the same:

"Whereas the Imperial Government of Japan has committed unprovoked acts of war against the Government and the people of the United States of America; therefore be it resolved by the Senate and House of Representatives of the United States of America in Congress assembled, that the state of war between the United States and the Imperial Government of Japan which has thus been thrust upon the United States is hereby formally declared; and the President is hereby authorized and directed to employ the entire naval and military forces of the United States and the resources of the Government to carry on war against the Imperial Government of Japan; and, to bring the conflict to a successful termination, all of the resources of the country are hereby pledged by the Congress of the United States."

It passed the Senate by a unanimous vote of 82 to 0 and was immediately referred to the House where it was introduced by Majority Leader McCormack. It passed the House by a vote of 388 to 1 and was duly signed by Vice President Wallace and by Speaker Rayburn for transmission to the White House. The one dissenting vote in the House was that of Rep. Rankin from Montana who had voted against war in 1917.

Retention of armed forces. The Senate today passed a bill retaining all officers and enlisted men in the armed forces of the United States for the duration of the war.

STATE DEPARTMENT – PRESS CONFERENCE. Secretary Hull in response to various questions advised the correspondents that Ambassador Litvinov had called this morning for the purpose of paying his respects that their conversation was brief and merely a prelude to those to follow involving matters of mutual interest and that the ceremony of presenting the Ambassador's credentials to the President may be attended to later today. He said that problems involving the return of our officials and nationals from Japan and the departure of Japanese officials and nationals from the United States were receiving attention but did not elaborate on the matter.

With regard to the possibility of a Pan American conference he indicated that the next scheduled Pan American conference was in 1943.

As for Nicaraguan and Costa Rican declaration of war on Japan, he said that this was a further exhibition of the increasingly fine spirit among the American nations.

Asked whether he had heard of the bombing in Hawaii while he met the Japanese envoys yesterday, he stated that he had received the representatives on the basis of the document they presented since he had not had an opportunity to confirm indefinite reports reaching him just before the arrival of the envoys.

With regard to our differentiating between the Koreans and the

Japanese the Secretary in reply to a question referred the correspondents to the Department of Justice.

RADIO BULLETIN. It has become necessary to reduce the length of the daily Radio Bulletin and for the time being at least to effect certain minor changes in its form of presentation.

ADD WHITE HOUSE – STATEMENTS. The following are the substance of various official statements made by the White House today:

American operations against the Japanese attacking force are still continuing. A number of Japanese planes and submarines have been destroyed. In Pearl Harbor 1 old American battleship was capsized and several other ships have been seriously damaged. 1 destroyer was blown up. Army and Navy fields were bombed with the destruction of several hangars and a large number of planes. A number of bombers have arrived safely from San Francisco. Guam, Wake and Midway Islands and Hong Kong have been attacked. Details are lacking. 200 Marines – all that remain in China – have been interned by the Japanese near Tientsin. Casualties on the island of Oahu are estimated at 3,000, nearly half of which are fatalities.

WAR RESOLUTION. President Roosevelt at 4:10 PM today signed the joint Congressional resolution declaring the existence of a state of war with the Japanese Empire.

American Embassy, Lima, Peru
December 8, 1941.

Radio Press News Lima Peru
December 8, 1941

RADIO PRESS NEWS LIMA PERU
DECEMBER 8 1941

STOCK MARKET :– Volume of trading 2,030,000 shares. London closed easy. First repercussions to the Japanese new lows since June 10, 1940, in the most active trading since May 22, 1940. Sharp declines in bonds with four United States government issues in new low ground and a sharp rise in commodities notably wheat and soy beans. Before the opening the Stock Market braced itself for a shock. But the preparations were unnecessary. Opening prices were down 1 to 3 points on blocs ranging to 6,500 shares in cottonseed oil and these executed without difficulty. Then United States Steel developed strength and the whole Stock Market moved up from the lows. Later when the war news seemed less favorable and President Roosevelt asked for a declaration of war quickly given by Congress, the Market declined again. Losses were extended to as much as ten points in Universal Pictures, first preferred. Most issues declined but there were a few strong spots, notably commodity stocks which traders believed would benefit through war demand. Several sugars made new highs and at the tops showed gains ranging to more than two points.

!!!!! 1500 DEAD IN HAWAII, ONE BATTLESHIP LOST !!!!!

WASHINGTON :– Casualties on the Hawaiian Island of Oahu in yesterday's Japanese air attack will amount to about 3000, including about 1500 fatalities, the White House announced today. The White House confirmed the loss in Pearl Harbor of "one old battleship" and a destroyer, which was blown up. Several other American ships were damaged and a large number of Army and Navy airplanes on Hawaiian fields were put out of commission the White House disclosed. It reported at the same time that American operations against Japan were being carried out on a large scale, resulting already in the destruction of "a number of Japanese planes and submarines." The White House statement said, "American operations against the Japanese attacking force in the neighborhood of the Hawaiian Islands are still continuing. A number of Japanese planes and submarines have been destroyed. The damage caused to our forces in Oahu in

236

yesterday's attack appears more serious than at first believed. In Pearl Harbor itself one old battleship has capsized and several other ships have been seriously damaged. One destroyer was blown up. Several other small ships were seriously hurt. Army and Navy fields were bombed with the resulting destruction of several hangars. A large number of planes were put out of commission. A number of bombers arrived safely from San Francisco during the engagement while it was under way. Reinforcements of planes are being rushed and repair work is under way on the shops, planes and ground facilities. Guam, Midway, Wake Islands and Honolulu have been attacked. Details of these attacks are lacking. Two hundred Marines, all remain in China, have been interned by the Japanese near Tientsin. The total number of casualties on the island of Oahu are not yet definitely known but in all probability will mount to about 3000. Nearly half of those are fatalities, the others being wounded. It seems clear from the report that many bombs were dropped in the city of Honolulu, resulting in a small number of casualties."

WASHINGTON :– President Roosevelt at 4:10 PM EST today signed the joint Congressional resolution declaring existence of a state of war with the Japanese Empire.

!!!!! CONGRESS PROCLAIMS WAR BETWEEN U.S. AND JAPANESE EMPIRE !!!!!

Congress today proclaimed existence of a state of war between the United States and the Japanese empire 33 minutes after the dramatic moment when President Roosevelt stood before a joint session to pledge that he will triumph, "so help us, God." Democracy was proving its right to a place in sun with a split second shift-over from peace to all-out war. The Senate acted first, adopting the resolution by a unanimous roll call vote of 82 to 0, within 21 minutes after the President had concluded his speech. The House voted immediately afterward and by 1:13 p.m. a majority of the House had voted "Aye." The final House vote was announced as 388 to 1. The lone negative vote was cast by Representative Jeannette Rankin, Republican of Montana, who also voted against entry into the World War in 1917 said this negation has no choice but to declare war on Japan. Only Miss Rankin and Representative Hoffman, Republican of Michigan, had remained seated when the House gave a standing ovation in response to President Roosevelt's solemn statement. I ask that the Congress declare that since the unprovoked and dastardly attack by Japan on Sunday, December 7, 1941, a state of war has existed between the United States and the Japanese Empire. In a staccato of short sentences, the President told where the Japanese had hit yesterday throughout the Pacific area and their representatives here had at the same time been continuing deceptive and false negotiations for maintenance of peace. And he said simply, he had ordered all measures be taken for our defense. No matter how long it may take us to overcome this premeditated invasion the American people in their righteous might will win through to absolute victory.

MANILA P.I. :– Press dispatches reported that 100 to 800 troops, sixty of them Americans, were killed or injured today when Japanese warplanes raided Iba, on the west coast of the island of Luzon, north of the Olangapo Naval Base.

NEW YORK :– Declarations of war since Japan's attack on the United States: Great Britain on Japan. Nicaragua on Japan. Canada on Japan. The Netherlands on Japan. Honduras on Japan. Costa Rica on Japan. Manchukuo on the United States. Free France on Japan. Haiti on Japan. Belgian Government in exile on Japan. Salvador on Japan. Imminent declarations: South Africa on Japan. Australia on Japan. Chungking on Japan, and the Axis. Cuba on Japan.

NEW YORK :– Japanese forces in the western Pacific have attacked the Australian mandated Ellis and Ocean Islands, both of which are rich in phosphate deposits, according to a British broadcast heard by the NBC.

SAN FRANCISCO :– The Singapore Radio, heard by a United Press listening post here today, reported that two American-built Hudson bombers operating off the northern Malayan coast had scored direct hits on two Japanese troop ships and another Hudson bomber had scored a direct hit on a barge loaded with Japanese soldiers.

NEW YORK :– The Japanese official radio claimed today that an agreement was reached with Thailand to allow passage of Japanese troops through that country. The broadcast was heard by the United Press listening post. Japanese Imperial headquarters reported that Japanese troops started to enter Thailand this afternoon, Radio Tokyo reported.

NEW YORK :– The Stock Exchange banned transactions in Japanese bonds. In London these issues had breaks ranging to 13 points despite the fact that Japanese assets frozen in London could continue debt services for a long time. Foreign bonds in New York weakened, including South American issues.

MANILA P.I. :– A United States Army spokesman said tonight that Japanese planes attacked Davao twice during the day. Unofficial reports circulated that three Japanese planes were shot down over Pampanga Province, north of Manila.

LONDON :– Prime Minister Churchill told Commons today that Britain had declared war against Japan and that insane ambition which is the root of evil must be extinguished, although the task will probably be long and hard. Churchill said the Cabinet had authorized him to announce the war declaration and disclosed that last night he talked by trans-Atlantic telephone with President Roosevelt.

CHICAGO :– Charles A. Lindbergh, isolationist spokesman, said today that we must meet war with Japan as united Americans.

NEW YORK :– The NBC correspondent at Manila reported that Japanese planes carried out a heavy attack on Fort McKinley and Nichols Airfield at Manila early Tuesday and started large fires. The Japanese bombers, despite a terrific curtain of anti-aircraft fire, touched off a huge fire, apparently in a gasoline dump.

SINGAPORE :– Major General Percival, British Commander in Chief for Malaya, reported today that the Japanese are believed to be in possession of southern Thailand and are fighting to retain their foothold on the beaches of northeast Malaya.

BUENOS AIRES :– Foreign Minister Enrique Ruiz Guinazu today conferred with Acting President Ramon S. Castillo after holding interviews with the American and Japanese Ambassadors. The Japanese Ambassador asked Ruiz Guinazu to define Argentina's position regarding the war. In a statement made later, Ruiz Guinazu said that Argentina will act automatically in accord with the circumstances, but will faithfully respect her international treaties.

RIO DE JANEIRO :– President Getulio Vargas of Brazil today affirmed his country's solidarity with the United States in the conflict with Japan. The Agencia Nacional, official new agency, issued a declaration, following a meeting of the President and Cabinet at Cattete Palace. The President in a meeting with the Cabinet decided unanimously to declare the solidarity of Brazil with the United States in accordance with continental compromises, the declaration said.

SINGAPORE :– British jungle fighters were reported locked in a struggle of annihilation tonight with Japanese invasion forces which landed in the Thailand border region of Malaya and struggled to push south toward the Singapore Naval Base.

HONG KONG :– Two air raids by Japanese planes on Hong Kong were beaten off by anti-aircraft fire today and damage was not important, a British command communiqué said.

ROME :– Newspapers today termed the Pacific war a gesture of Axis solidarity on the part of Japan. There was no official indication whether the Tri Partite Pact would be invoked to send Italy and Germany to Japan's aid, but competent quarters expressed complete solidarity and sympathy with Japan. They denounced President Roosevelt's war mongering and said the Japanese declaration of war had made on enormous impression on the Italian people.

PANAMA CITY :– Dozens of deadly P-40 pursuit planes droned back and forth across the Isthmus of Panama today, protecting the Panama Canal from attack. Lieut. General Frank Andrews announced that war plans had been put into effect and that all commercial permits to fly over the Canal Zone had been rescinded.

OTTAWA :– Canada was formally at war against Japan today because the Island Empire had wantonly and treacherously attacked British and American territory and forces.

BERLIN :– A Nazi military spokesman said tonight that Germany has abandoned attempts to capture Moscow for this winter. The military spokesman said that large scale operations by Germany on the eastern front have been ended for this winter. Moscow, he said, will not be captured before spring. He said the early start of the Russian winter caused the Germans to make this decision.

WASHINGTON :– The White House tonight accused Germany of having done all it could to push Japan into the war. An official statement said: "Obviously Germany did all it could to push Japan into the war. It was the German hope that if the United States and Japan could be pushed into war that such a conflict would put an end to the Lend Lease program. As usual the wish was father to the thought behind the broadcasts and announcements emanating from Germany with relation to the war and the Lend Lease program. That such German broadcasts and announcements are continuously and completely one hundred per-cent inaccurate is shown by the fact that the Lend Lease program is and will continue in full operation."

End of RCA Special News Bulletin.
American Embassy, Lima, Peru
December 8 1941.

* Note: The above document has been transcribed as in the original, misspellings have been not been corrected.

Vernon dancing with Major-General Roy Geiger's grand-daughter, Melanie Johnson,
the flower girl for LaVerne's wedding on December 21, 1954 in Hawaii.
LaVerne had been the flower girl at Melanie's parents', Joyce Geiger and
Bob Johnson, wedding in 1937 at Quantico, Virginia.
This picture was published across the country by the Associated Press.

Major General (then Brigadier General) Smedley D. Butler and Major General John A. Lejeune (then 13th Commandant of the Marine Corps) watching the Marines play baseball at Washington, D.C. in 1922. Vernon served under each of them. He viewed Lejeune as one of the best of the Corps Commandants and Butler as a combination of George Armstrong Custer and George Patton of World War II fame.

Vernon's campfire stories never dealt with war, but always with his love of flying

On command of the First Marine Air Wing in Korea. "Although higher rank and greater responsibility were to come my way later, the apex of my military career was reached in Korea."

Not even its hard-won experience during the long, vicious Haitian Campaign could prepare the U.S. Marine Corps for the five-year Second Nicaraguan Campaign, because Nicaragua was five times the size of Haiti. A new weapon had to be found, and this it was that the Marine Corps' air-ground team came into being.

Six DH-4 biplanes (*top, left*), the first Marine aircraft to be committed to the Nicaraguan Campaign, were shipped in crates to Managua where they operated out of an unimproved cow pasture (*center photo*). Other Marine airplanes employed were the two-passenger OL-8 amphibians (*top, right*), the OC-15 (*above, left*), which arrived in 1929, and the O2Us (*above, right*), which replaced the DH-4s.

Capt. Vernon E. Megee, Executive Officer of VF-9M

To order additional copies of this book,
contact the publisher:

Atriad Press LLC
13820 Methuen Green
Dallas, TX 75240
(972) 671-0002
www.atriadpress.com

www.ingramcontent.com/pod-product-compliance
Lightning Source LLC
Chambersburg PA
CBHW060044100426
42742CB00014B/2695

* 9 781933 177281 *